PÂTE DE VERRE
AND
KILN CASTING
OF
GLASS

By Jim Kervin and Dan Fenton

© 2000 James Kervin and Dan Fenton

Pâte de Verre and Kiln Casting of Glass

by James Kervin
and
Dan Fenton

Published by:

GlassWear Studios
1197 Sherry Way
Livermore, CA 94550
(925)443-9139 Fax (925) 292-8648

All rights reserved. No part of this work covered by copyright hereon may be reproduced or used in any form or by any means - graphic, electronic, or mechanical, including photocopying, recording, taping, or information storage and retrieval systems — without permission of the publisher, except for the inclusion of brief quotations in a review.

Copyright © 1997
First Printing 1997
Second completely revised edition 2000

Cataloging Data
Kervin, James E., 1950-
Fenton, Daniel M., 1950-
    Pâte de Verre and Kiln Casting of Glass
        by James E. Kervin and Daniel Fenton 292p
    Includes bibliographical references and index
    ISBN 0-9651458-3-2
      1. Glass blowing and working. 1. Title
TP859.K   2000

Library of Congress Control Number 00-092629

Front cover: "Maestro en Rouge" by Seth Randal
                Pâte de cristal, 1991
                10" high by 23" in diameter
                Photographed by Roger Schreiber
                Collection of Ginny Ruffner

Preface

This cooperative effort was established to develop a text on the complex glassworking technique of pâte de verre or kiln casting of glass. The process allows a small glass studio artist who possesses a kiln to push their artistic envelope into the fascinating world of three-dimensional sculptural glass. This book, as would be expected of any book by Dan or Jim, is a very comprehensive presentation of the subject ranging from initial development of the model to final cleaning of the glass casting. It is loaded with detailed tips on the process as well as information pertinent to kilnworking of glass in general.

Dan Fenton is an internationally known glass artist who has been working in the field of glass since the sixties and with kiln techniques since the late seventies. He has long been known as a walking storehouse of detailed technical information on many small studio glassworking techniques and travels the country teaching pâte de verre as well as teaching it at his studio in Oakland, CA. Many other glass art subjects are also taught there. They include: stained glass, sandblasting, fusing, glass painting, dale de verre, neon, enamel work, beadmaking, etc. Dan is well known in the glass community for his many articles, willingness to share his vast knowledge and his own unique sense of humor. In fact, this work builds heavily on a series of articles he wrote on the subject of pâte de verre and kiln casting of glass for Glass Art Magazine.

Jim Kervin is a relative newcomer to glass, having only fifteen years of experience in stained glass, ten in kilnworking and eight in lampworking. His strength is that with a Ph.D. in engineering, he tends to approach a subject from an orderly and scientific viewpoint. He wants to know how and understand why things work the way they do. Some of you may already know him from his book on glass beadmaking entitled *More Than You Ever Wanted To Know About Glass Beadmaking*. He was interested in developing this book to make information on kiln casting of glass more available to the average glass artist. He had previously found himself frustrated when searching for information on this subject and wanted to make sure that the information was easily available to any budding artist seeking it.

Together, we have worked to compile all the information on the subject of pâte de verre and kiln casting of glass that we think the average glass artist could ever want. We hope that this book fulfills your needs for information on the subject and facilitates you in delving deeper into the possibilities available using this fascinating technique.

We wish to thank the many artists who have been so sharing in information on the techniques that they use in making pâte de verre and kiln cast glass. In particular, we would like to thank Mark Abildgaard, Anna Boothe, Ruth Brockmann, Linda Ethier, Newy Fagan, Larry Fielder, Mary Fox, P. J. Friend, Robin Grebe, Lucartha Kohler. Boyce Lundstrom, Donna Milliron, Charles Miner, Seth Randal, Alice Rogan-Nelson, David Ruth, Kathleen Stevens, Janusz Walentynowicz and Mary Francis Wawrytko.

It should be mentioned that any discussion of particular equipment or supplies in this book does not constitute a recommendation. They are just some of those that we use in our work or know about from others. Results achievable with them may vary from individual to individual depending upon their skill level and the application technique; so no warranty is implied for these products. Instead we supply this information so that you, as an educated consumer, can make your own choices.

Because of the nature of this art form, injuries can occur and you, the user of these products, have the sole responsibility for your safety as well as for other individuals that could be hurt by your actions. Neither we, nor the equipment manufacturers, are responsible for any injury resulting from equipment misuse. You need to learn as much about how your equipment operates as possible, read and follow all manufacturer recommendations, and take classes from qualified instructors. With this type of background, you will be able to safely enjoy pâte de verre and kiln casting of glass.

This book was designed to provide as complete an information source as possible regarding the subject of kiln casting of glass. Although this topic is presented in great depth, it is impossible to cover everything. You are urged to read all available material on the subject. Check out the many references listed herein to supplement this text. Be aware that, even though we have tried to provide as complete and accurate a source book as possible, **there may be some mistakes** present both typographical and in content. Therefore, use it as a guide and if something seems wrong, realize that it might just be so.

This manual was written both to educate and to entertain the reader. The authors shall have neither liability nor responsibility to any person or entity with respect to any loss or damage caused, or alleged to be caused, directly or indirectly by the information contained in this book. **If you do not wish to be bound by these restrictions, you may return this book with a copy of your receipt to the publisher for a full refund.**

Jim Kervin and Dan Fenton

Table of Contents

PREFACE 3

TABLE OF CONTENTS 5

INTRODUCTION 7

BASIC KILN CASTING PROCESS 11

GLASS 25

MODELING 43

REPLICA CASTING 63

INVESTMENT MATERIALS AND FORMULATIONS 73

MOLD CONSTRUCTION AND PROCESSING 85

ADVANCED CASTING TECHNIQUES 109

KILN PROCEDURES 121

FINISHING TECHNIQUES 137

SAFETY 155

GLOSSARY 169

SUPPLIERS AND MANUFACTURER SOURCES 171

REFERENCES 175

INDEX 179

© 2000 James Kervin and Dan Fenton

Introduction

Pâte de Verre is an ancient glassworking technique that was rediscovered in the art nouveau period of the late 1800's by Henry Isadore Cesar Cros. The term pâte de verre can be translated literally from French as "paste of glass." The technique produces solid glass sculptural objects with a rich translucent quality. The process was utilized extensively by the ancient Egyptians to make glass amulets, scarabs and gems.

Pâte de Verre techniques can be used to produce a wide range of objects from solid cast sculptures to hollow vessels. The technique can be used to produce single colored glass objects or ones with great surface detail and variation in color. The glass surface of a pâte de verre work is most often matte, but it may also have a wet rough texture. Above all, it results in a work that begs to be held and examined.

Figure 1. Pâte de verre bas-relied plaque
Artist: Henri Cros 1886 France
Courtesy of the Corning Museum of Glass.

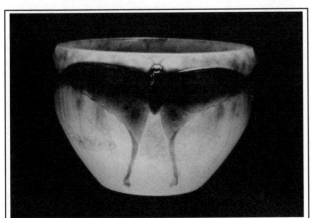

Figure 2. Pâte de verre butterfly bowl.
Artist: Gabriel Argy Rousseau about 1915.
Courtesy of The Corning Museum of Glass.

The basic process

You may have seen this type of work before and wondered how it was made. The pâte de verre process is actually quite involved. It starts with the making of an original model of some material. Depending on the complexity of the model and your desire for replicas of it, you might next make a master mold of the model. With a master mold, you can cast as many wax replicas of the model as you

Figure 3. "Sirene au Poisson" pâte de verre sculpture France circa 1925-1935
Courtesy of the Corning Museum of Glass.

Although the whole process may sound complicated beyond your wildest expectations, don't let it scare you. If we can do it, so can you. Just go through the steps carefully the first couple of times and you will get the hang of it. Don't expect to create a masterpiece your first time out, but be assured that you will get better with practice.

We also suggest that you start off small before you proceed on to larger masterpieces. Besides being easier, doing smaller pieces will help conserve on materials, help keep the initial messes down, and prevent the vise grips of major failure from clamping down on your mental health.

Kiln casting versus pâte de verre

The term, pâte de verre, is loosely used by many glass artists today, perhaps because of the romantic sound of the French term. They apply it to any glass artwork created by kiln casting of glass frit in a refractory mold regardless of the size of the frit.

Such "chunk de verre" is not representative of what the old masters produced. It produces objects that look like they are made from marble. It also lacks the fine detail and the crystalline quality of the original work but does produce art objects with their own beauty. Such work should probably be differentiated from the traditional pâte de verre work by use of a different term such as frit or kiln casting to avoid confusion. We try to make that distinction in this book.

desire. The replica, or original model, is next used to cast a refractory mold.

After setup of the investment and removal of the model, you air dry the mold and then fire it in your kiln to burn out any residual organic materials. Then a fine glass paste is packed or painted into the mold in a subtle blending of thin multi-colored layers. Of course, you will have to make this paste yourself from colored glass or clear glass that you color and then frit (make into glass granules).

After drying, the thin layers of glass paste are fired in your kiln until they are completely fused together. Successive layers of frit can be added and fused until the piece is complete. Once a piece is fired and annealed, the mold is broken away to reveal the completed object of art. Lastly you clean the casting and sit back for a well-deserved rest. Some artists though, may continue on and reinvest a casting or glue pieces together as part of a larger work.

Figure 4. Work by Fredrick Carter
Courtesy of the Corning Museum of Glass.

© 2000 James Kervin and Dan Fenton

The path forward

This process draws a lot on tools, supplies and skills that were developed by other artisans; such as: jewelers, sculptors, metalsmiths, potters, fusers, and glassblowers. You will learn to work with many materials to sculpt your original models. You will learn lost wax casting techniques. You will even learn to mix you own glass colors.

You will find that many of the tools and materials that you will be using in pâte de verre are not available at your local stained glass retail store. Part of the adventure in starting out on this process will be in seeking out sources for these new tools and supplies. As a result of this search, you will make more than one visit to any of a number of new locations seeking out these materials. They may include: hardware stores, jewelers' supply stores, ceramic supply stores, flea markets, art supply stores, and even your local supermarket.

So sit back and let's embark on this new glass journey. Together we will marvel at the intricacies we see along the way, gawk at the beauty of the work, and envision new vistas for the future.

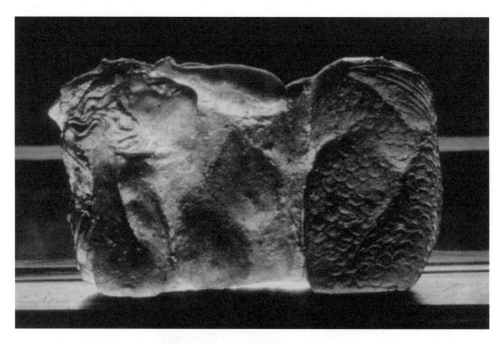

"Mermaid" 1993
"Chunk de Verre"
Size: 14" x 7" x 2"
Artist: Kathy Bradford
Photo by: the artist

Basic Kiln Casting Process

We were originally a little undecided on how to organize this book. Should we start right in with a step by step description of the pâte de verre process, presenting all the details along the way or just give you an overview of the process and then go into more detail on the some topics later. But we have found that sometimes, even we are not always interested in the details of a new technique until we have first developed an understanding of the basic process.

So after much philosophical discussion, we decided that it was better to present you with the basics of the process itself here and then in later chapters to go back and wade our way through all those esoteric details. At that point, you will better understand why they are important and how they all fit together.

So let's get down to basics.

Glass types

Choosing the type of glass to use in your work is the first decision that you will have to make. The choice of glass dictates the processing temperatures, which in turn has a profound effect on the results that you will be able to achieve in a casting. The higher the fusing temperature of the glass, the more refractory the investment mixture has to be and the more likely it is to develop cracks.

As will be discussed later in subsequent chapters, higher temperature refractory investment materials tend to retain less detail or can be so strong as to endanger the piece when demolding. These factors have to be weighed against other factors like cost, availability, and compatibility of the materials that you use.

For now let's just assume that you are making a casting from a single piece of scrap soda-lime glass left over from your stained glass career which you are now going to break into pieces. Try to use a light, transparent colored piece of glass so that you will be able to see through it. If the color density is too great, the body of the work will look dark and uninviting.

Frit preparation

The opacity of a pâte de verre work besides being a function of color density, is also the result of fine air bubbles trapped between individual grains of the glass frit. In kiln casting, you do not usually get the glass hot enough for bubbles to rise to the surface of the glass. This result is contrasted to a typical molten glass casting process where the glass has been fined (debubbled) prior to pouring it into the mold. The number and size of the air bubbles trapped in the finished pâte de verre piece is dictated by the size of the glass frit used in its manufacture.

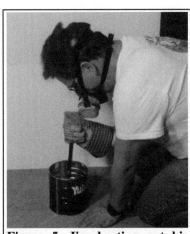

Figure 5. Jim beating out his frustrations making frit.

Pieces made with frit the size of flour will end up looking more like alabaster than glass. Use of larger frit, the size of sugar, will result in a piece with more translucence but still with a definite opacity. Larger particles yet will continue to increase the transparency of the piece until you reach the ultimate limit of using sheets of glass. Here the final result looks essentially the same as the initial glass except for a few bubbles, which may get trapped at the interfaces of the sheets. These bubbles could be minimized even further by orienting the sheet glass pieces vertically in the mold.

For this first kiln casting project, we would suggest that you crush up part of a single sheet of glass from your stained glass stock. Do this by wrapping it up in a piece of newspaper, laying it on your garage floor, and working out your frustrations on it with a hammer. Or you can stick it in a coffee can as Jim is doing in Figure 5. This is an especially fitting end for one of those pieces of stained glass that just never cut like it should. Their type has tortured us all many times in the past.

After crushing the glass, you need to wash the frit to remove any dirt and organic material. This will also get rid of a lot of the really fine glass dust that some artists don't like and that is an inhalation hazard. Then spread the frit out to dry on some clean newspaper. You will want to run a magnet over it after it is dry to pull out any iron filings from the hammer. (This is normally done on any frit produced by mechanical methods.)

Lastly you may want to divide the frit into different size ranges. Do this by running your frit through various sized kitchen sieves or wire meshes that can be purchased from a ceramic supply store. This is what Jim is doing in Figure 6. Now, set the frit aside for later.

Figure 6. Sizing your frit using wire meshes.

Model making

The first basic step in the kiln casting process is to develop a model of the work you want to cast in glass. This is where we let the artist, or the little kid in all of us, out of their cage to develop the theme for our work. The model, that he or she develops, is the original object that we will transform through the magic of the kiln casting process into glass.

You can develop this model by working with any of a number of materials, the choice of which material you use depends on whether you want to go directly into casting your refractory mold from your model or you want to add an intermediate step of making a wax reproduction. For now, we will deal with the former and save the latter for the chapter on replica casting.

Next we need a palette on which to construct our model. A good palette choice on which to work your model is a piece of ¼-inch thick plate glass. It is strong and easy to clean up afterward with a solvent like alcohol. Besides we are sure that most of you probably have some plate glass lying around. Dan prefers to use cardboard for reasons that will be explained later.

One of the things that you have to consider when making the original model is that the glass shrinks in volume by 30 to 50% in its transformation from frit to a fused solid piece. Therefore, you have to plan for how you are going to add the extra glass needed to compensate for that shrinkage.

If your piece is to have a large flat back like a mask or a medallion, you could just pile a lot of frit in the large opening at the top of your mold as shown in Figure 7. Then later, after soaking at fusing temperature, you could peek into the kiln to see if you need to add more glass. If so, you turn off your kiln; don your protective clothing, gloves and glasses; and pour some extra frit from a scoop onto the pile.

If you do not especially like to singe your eye lashes digging around in a hot kiln, you may want to plan on incorporating a reservoir into your mold to hold the

Figure 7. Frit mounded high in an open face mold.

extra frit that will be needed to fill the inherent void volume of the frit as it fuses. This reservoir can be incorporated in the form of a base that you retain as a part of the finished piece or as a temporary funnel that you grind off as desired later. This base or funnel is attached directly to your plate glass palette during modeling.

There are many materials from which an original model can be developed. They include:
- **Wax** is the traditional used by artists like jewelers for casting of metals. The beauty of this material is that the model is easily removed after casting the mold by melting out the wax.
- **Combustible objects** are another good choice to use as a model. In this case, the object can be removed from the mold by burning it out when you cure the mold. This can be a smelly and smoky process so be sure that your kiln is well vented.
- **Clay** is a good choice if you are making a large flat object like a medallion with a minimum of under cuts. After investing the mold around the clay, the clay can just be pulled out and the mold cleaned with water to remove any clay residue.

Chose whichever medium seems right for you and get started on your model.

Mold construction

Upon completion of your original model, it is time to construct a plaster-based mold around it. In this section, we discuss the basics of the process for making single-part waste molds, which are used once and destroyed in demolding. Later on, in the chapter on investments and the one on advanced mold construction, we will delve deeper into the materials that go into an investment mix, give some sample investment formulations, and discuss more advanced mold construction techniques.

In the basic process, the first step is to completely clean the plate glass palette on which the original was made. This is done in order to be able to anchor down a mold frame to the glass palette so that it will not leak. In this frame, we will cast the refractory investment mold around the original model.

So let's start by planning out the size of the investment mold to make around the original model. The mold, as illustrated in Figure 8 for the case of an artichoke casting with an incorporated reservoir base, should typically be at least ½ inch thick (t) and at most about two inches thick. This holds for the base as well as the sides of the mold.

How thick you make your mold depends a lot on which exact process you will be using. If you are using traditional pâte de verre techniques as practiced by the old French pâte de verre masters where you will fire the mold multiple times as you build up additional thickness of a piece, you may want the mold to be on the thicker end of that range. If instead, you are doing a one step casting of an object with a low firing-temperature glass, you can probably get away with making your mold on the lower end of the thickness range. Also as your model gets larger, your mold will need to be thicker and be reinforced to resist the outward pressure from the molten glass.

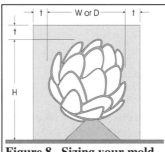

Figure 8. Sizing your mold.

Be sure to plan for shrinkage in volume as the frit consolidates. If you really want about a half-inch thick medallion, then make the model three-quarters to an inch thick. If you want a reservoir base that you are sure will completely fill a mold, then the volume of the reservoir should be approximately the same as the model.

Making a mold frame

To cast your mold, you need to first construct a mold frame or box around your model in which to cast it. Jewelers usually use a cylindrical metal frame referred to as a flask for this. Here we will present a construction of a simple cardboard mold frame. Frames from cardboard are good for small projects using up to about 5 to 20 lb. of investment material. For something larger than this, you need a stronger mold frame material.

Cardboard is good choice because it is easy to work and easy to come by. Avoid any of the waxed cardboard like on fruit boxes because the hot glue will not stick to it. Cardboard is easily cut with a standard utility knife. Dan's favorite is a Spyderco single serrated edge knife, but don't go running out to buy one because they are fairly expensive. The bottom of the mold frame will be glued to the plate glass or cardboard palette on which your model is mounted.

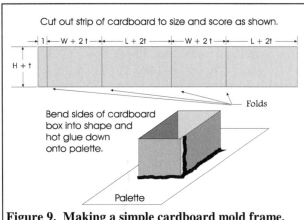

Figure 9. Making a simple cardboard mold frame.

To determine the dimensions of your frame, both length and width, envision the mold on the palette of the decided upon thickness around the model as was illustrated in Figure 8. The cardboard frame will have sides that are ½ to 2 inches higher than your model is high (again based upon the desired mold thickness, t) and sides at least 1 inch longer than the model (2t).

Therefore, cut out a strip of cardboard as is illustrated in Figure 9 that is H+t wide and about one inch longer than twice the sum of the length (L+2t) and width (W+2t) of your intended mold. You will use the extra one-inch to make a flap to glue onto the first side to close the box. Fold the cardboard where each of the corners will be. Folding the sides can be made easier by scoring fold lines on the backside of the cardboard prior to folding.

Use a hot glue gun to attach the ends of the cardboard mold frame together and down to the palette. Glue guns are easy to use. Choose the trigger type of glue gun available from any hardware store because it makes feeding of the glue sticks easier. Don't worry about getting a trigger lock though because these guns are entirely non-lethal. The guns are an item that can also often be found at a flea market for less than $3 and will almost always work okay.

Glue sticks are a consumable item that you will find yourself going through like crazy. Each mold frame may require up to a half dozen of small ones. Some major hardware stores sell them in bulk and, if you can find them, buy the long sticks so that you don't have to reload the gun as often. The clear stick glue is stronger but the white "caulk" sticks set up faster.

Be prepared to get an occasional "ouchie" and to exercise your vocabulary because in the frenzy of creativity, some of the hot glue will invariably get on your hands. It is never fatal and only rarely does it result in burns. If you are really sensitive about your lily-soft paws, you might want to wear dishwashing gloves while gluing. Be sure to clean up all the wispy glue strands after you are done.

Figure 11. Placing the mold frame around a model.

Getting good glue joints between the palette and the mold frame is where it is critical that you did a good job cleaning your glass palette before attaching the cardboard sides. Nothing is less amusing than having investment leak out under a side of your mold frame because of a poor palette-cleaning job. This is why Dan prefers using cardboard for a palette—it bonds better. Jim prefers glass because it is stronger and allows picking up the filled mold frame and moving it out of the way.

If the mold will use more than about 5 lb. of investment, you may need to provide external reinforcement to the cardboard to help prevent it from bulging out and tearing away from the hot glue. Reinforcement can be in the form of cardboard buttresses or just piling bricks around the sides of the mold frame. (In case you are wondering, buttressing was an architectural technique used in building Medieval cathedrals and here refers to adding triangular pieces of cardboard attached perpendicular to a mold frame wall and attached with

Figure 10. Hot gluing the mold frame down to the palette.

Basic Kiln Casting Process   15

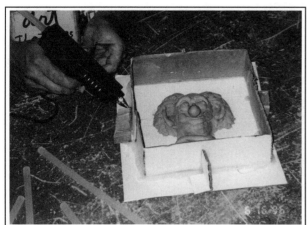

Figure 12. Reinforcing a cardboard mold frame.

hot glue both to the wall and the palette as shown in Figure 12.)

Once your mold frame is fully constructed around the model, lightly coat the palette and the inside of the mold frame with a separator compound. This should be some hydrocarbon liquid of your choice. It is easiest if this coating can be applied as a spray. Products such as liquid dishwashing detergents, cooking pan coatings (Pam), light machine oil sprays (WD-40), or green soap liquids are suggested.

Depending on what you used for your original model and how you plan to remove it from the mold, you may want to coat it also. For materials that you are going to burn or melt out, this is not necessary. If you choose to use separator on the original model, be aware that some surface detail will be lost. So paint it on thin and get rid of any bubbles. The separator compound should assist in reducing any bubble formation in the investment at the model surface by reducing the investment's surface tension.

Mixing the investment

Now it is time to mix up a batch of investment. Most investment recipes are plaster-based and mixing them can take some practice. For this exercise, we suggest that you use what will be called investment mix # 1. It consists of a fifty-fifty mix by weight of Hydrocal plaster and 200-mesh silica flour. Wear a respirator because the silica flour dust is hazardous. Also consider wearing splash goggles to keep it out of your eyes and gloves if you have sensitive skin.

Mixing plaster-based investment is often assumed to be elementary but there are really many different aspects that need to be addressed to do it right. Don't expect to necessarily get it right the first time. With a little practice though, you will soon become a professional.

Try to mix enough investment to completely fill your mold frame in one pour, as this will make the strongest possible mold. If this is not possible, allow the previous batch of investment to set up before adding the next batch.

In mixing any plaster-based material by feel, as we will describe here, there are a number of suggested rules that should be followed to achieve optimum results.

Choose the proper sized mixing vessel for the amount of investment you are making up. It should not start off filled more than one half full of water since the volume of the final mixed investment will be roughly twice the volume of the starting water. Better yet, use a vessel that will start off only about one-third full so it will not overflow during mixing.

Make sure that your water is clean. In dirty water, new investment will coat the old investment to form clumps. Some people also swear that letting your water sit in a bucket for a few hours before use to be beneficial to final investment consistency. This allows a lot of the junk and gas that the city puts into your water to come out. (They also say that this also

Figure 13. Sifting the investment into mixing bowl.

© 2000 James Kervin and Dan Fenton

makes their water taste better.) If you want to go to real extremes you could use distilled water.

Always use cool, about 70°F, water to make up your mix. Maximum plaster solubility occurs between 70° and 100°F. Water that is too hot will cause investment to set faster. For those real control freaks out there, you can purchase an attachment for your sink from a darkroom equipment supplier to measure faucet temperature.

Keep your mixing bucket and tools clean. The presence of any old hardened investment will also cause the new investment to set up faster, in as much as one half the expected setup time.

Add the plaster-based investment mixture to the water, not the water to the investment. Don't just dump it all in at once. Instead, slowly sift the dry ingredients out onto the surface of the water as illustrated in Figure 13. An old flour sifter works okay for this purpose. Slow sifting of the investment onto the water avoids excessive air entrapment by the large surface area of the plaster particles. As the mixture settles onto the water, it will absorb enough water to displace the trapped surface air and sink.

Some mix formulations give suggested water to mix ratios. If so, add the dry mix until you reach that optimum ratio. You will need to preweigh your dry and wet components ahead of time to get this right or mix it on a scale.

If mixing by feel, keep sifting investment mix onto the water until it "peaks" as shown in Figure 14. This is where it no longer sinks and the surface develops slightly dry islands or peaks as illustrated.

Figure 14. Investment peaking on the surface.

This indicates a good dry ingredient to water ratio. Stop adding plaster-based investment mix at this point. Never stir the investment as you add it because that will mess up the cure and it will throw off the peaking measurement.

Figure 15. Hand stirring the investment.

Next, allow the mixture to sit undisturbed and rest for about three minutes. This time can be varied from as short as one minute to as long as five minutes. This time period is called "slaking" and is necessary for the investment to set properly. During this time, the plaster in the investment is becoming uniformly wetted and is beginning to chemically adsorb the water. You may want to tap the edge of the container during this time to get rid of any bubbles.

Now you hand stir the investment mixture being careful to avoid whipping any air into it. If you encounter any lumps as you mix the investment, this may indicate that you added the plaster to the water too quickly or that not enough time was allowed for slaking. Of course, there is also the possibility that your investment was lumpy to begin with. Check it out. If there are lumps in your dry mix, you can screen them out next time. In such a case, you should also consider just getting a whole new batch of investment.

Slowly stir the investment mix from the bottom to the top allowing any trapped air to rise to the surface. The mixed investment should appear to be creamy at this point, not thin and watery but slightly opaque on the stirrer. Do not overstir. First, because all the time that you are stirring, the mixture is starting to set up and second, because the more that you stir,

© 2000 James Kervin and Dan Fenton

Basic Kiln Casting Process 17

the faster it seems to set up. In fact you may want to undermix the batch to allow longer setting times.

Pouring the investment

After mixing, apply the investment mix to the model by slowly pouring it into the mold frame. Try not to trap bubbles. It works best if you pour the investment into an empty corner of the mold frame and allow it to flow around your model until you fill the mold frame to the rim.

Afterward vibrate the mold gently by tapping on the edge of the palette or shaking the worktable to free any trapped air and allow it to rise to the surface. This is where Jim sometimes has trouble with the mold frame breaking free from his glass palette and allowing investment to flow out from underneath. He then finds himself holding down the corners of the frame for the next half-hour. So try to make sure something interesting is on the television or the radio, just in case.

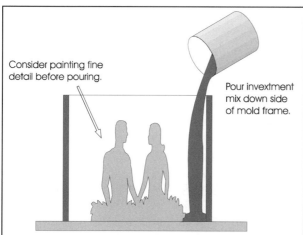

Figure 16. Pour investment into corner of the mold frame and let it fill up around mold released model.

Bubbles can also be broken free from the model by careful stirring of the poured material, assuming that you can do so without touching the model.

It is best to mix slightly more investment than you may think is necessary for the job rather than not enough. Otherwise if you have to mix more, the original batch will already be setting up just as the second batch is finished mixing and ready to pour. This will prevent the two layers from intermingling and they will then usually separate later.

It is a good idea to have some small molds ready to pour just in case you mix way too much investment. That way you will not waste any of it. Try to have your pour completed within about 10 minutes of initial mixing for best results.

Figure 17. Brushing on first layer of investment.

If you have a lot of detail that you are trying to capture from your model, you may want to try to paint a thin initial layer of investment onto it with a soft brush prior to pouring the bulk of the investment. Make sure that you fill any indentations that may trap air as you pour your investment. If you are fast enough, you may even be able to brush on your first layer onto the model and then fill the rest of the mold frame, all with the same batch of investment. If not, just take your time and let the painted layer set up before pouring the bulk of the mold.

If you end up needing to add a second pour of investment material, let the first pour set completely first. Then lightly scratch the top of the first pour to increase its surface area. This gives the next batch of investment material something to grab on to. Also saturate the material from the first pour with water or it will steal water out of the second batch, preventing it from setting properly.

Now that you are done with your pour, don't forget to clean your investment-coated tools and bucket with water right away. Start by pouring off all of the extra investment into some disposable trash container such as a plastic bag lining the inside of a cardboard box as shown in Figure 18. Let it set up before you try to get rid of it. Otherwise the bag can rip and make a mess.

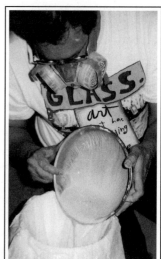

Figure 18. Disposing of excess investment.

© 2000 James Kervin and Dan Fenton

Do not clean your equipment in your sink or other unmentionable plumbing fixtures. Never dump plaster or investment down a drain, unless you don't like your landlord and you are moving real soon. The investment particles are fairly heavy and will settle out in your pipes. There the plaster-based investment will even set up underwater and harden. This practice could eventually lead to complete blockage of your drainpipes creating problems for everyone but your plumber, who probably could use the work.

Figure 19. Screeding the investment pour.

Instead, always have another large plastic container, like a dishwashing pan, of water available in which to rinse everything off. Let the heavier plaster particles sink to the bottom of the plastic container. Then dispose of the water at top in an appropriate manner, like giving a tree a drink. Discard the plaster sludge at the bottom into your plastic bag for disposal. If any investment dries in the plastic container before you can clean it out, don't worry it won't stick to the plastic. Just kick the side of the plastic container and the plaster will come loose.

When first poured, investment will look shiny and have a slight water film. Then as you watch, the shiny appearance will go away as the water is bound up by the investment. This phenomenon is known as glossing off and should occur within about 12 minutes after thoroughly mixing plaster-based investment. As the shine goes away and the investment starts to get mushy, you will want to screed what will later be the bottom of your mold, but is now the exposed top of your pour.

Screeding consists of lightly drawing a flat instrument like a metal ruler across the top of your pour as shown in Figure 19 to form a nice flat bottom for your mold to sit on the kiln shelf. This is done by making a number of light passes to avoid ripping up the smooth investment surface and is complete when the investment is flush with the top of the mold frame. Clean the investment off your screeding instrument after each pass to avoid tearing of the investment surface.

The results of investment mixes may vary, but most should start to harden up in about 20 to 30 minutes. If not, you may have gotten a bad batch of material or have somehow done something drastically wrong.

Once the investment in the mold has initially set up, remove the mold frame from around mold and any hot glue stuck to the palette. Then slide the mold off of the palette. Setup for plaster-based molds will generally take about two hours to for the investment to develop the proper set of chemical bonds between the plaster crystals.

Next you should scrape all the edges of the mold and any visible joint locations of the mold frame with a sharp knife as illustrated in Figure 20. You are trying to remove any sharp edges or divots from the surface of the mold. These may serve as points for initiation of cracks during drying or curing and their removal generally results in reduced cracking of the mold. Many people tell us that this step is unnecessary but that is what we learned when we started out and we feel honor bound to pass it on to you. You decide if you think it is necessary or not.

Curing the mold

The investment mold around the model is what holds the glass frit and eventually goes into the kiln. So it will have to be carefully cured. After it has completely set up, it is time to remove your model. Since most plaster-based investment mixes expand slightly during curing, a solid model if properly designed should be easy to remove. Assuming that

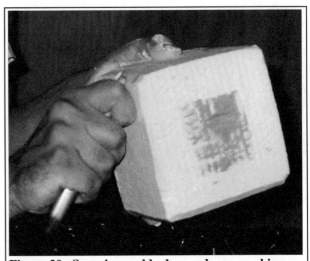

Figure 20. Scraping mold edges reduces cracking.

Basic Kiln Casting Process 19

you having plenty of draft and don't have undercuts as is shown in Figure 21.

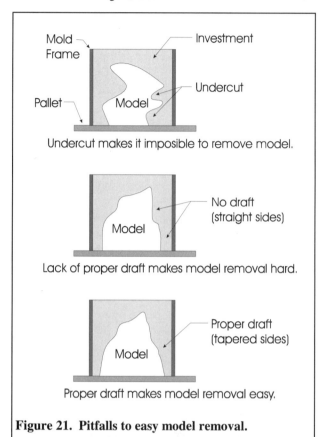

Figure 21. Pitfalls to easy model removal.

You may be able pry the model out of the mold with a sharp tool around the edges of the model. Injecting compressed air under your model is also helpful in breaking the suction between it and the mold caused by the separator compound. A quick pull or jet of air may not always be sufficient to break this vacuum but a slow steady pull might. An easy way to apply a slow steady pull is to hang the whole thing by the model a small distance off the table and let gravity do the work for you.

If your model is made of something like clay, you can carve out the bulk of it from the mold and wash out any residual with warm water. Here undercuts and draft are not of concern. All the clay must be removed or it will contaminate your final work.

With a wax model, you can melt or steam it out after your mold has set. To do this, support your mold on a screen or rack over a pan of water in your kiln or on a hot plate. The pan should be big enough to catch all your wax. An aluminum pie pan works well for this. Fill the pan with water and slowly bring the kiln up to about 225°F. As the mold gets warm, most of the wax will melt and drip out into the water.

If you are really frugal, this wax can be reused later but we do not necessarily advise it since the wax can pick up pieces of investment and get gritty. Don't let all the water evaporate, otherwise the wax will start to vaporize and create an explosive hazard. If you need to add more water, turn off your kiln and pour some into the pan.

After you have melted out all the wax you think possible, you may have to burn out the rest. This wax residue is typically burned out to avoid staining your piece with carbon or having a reducing atmosphere in the kiln while casting your piece. A reducing atmosphere can cause other problems, such as making lead precipitate out of lead-based glasses. Once all the wax is burned out of the mold turn off the kiln and allow it to cool before removing the mold and the pan of water.

For best results, the mold should be thoroughly air dried before putting it in the kiln for burnout and curing. Air-drying reduces the amount of water in the mold so that it does not turn into steam when later heated in the kiln. Expanding steam stresses the investment matrix of the mold, thus weakening it. Air-drying can be done in a number of ways.

The easiest way is just to set it aside for about a week and let it air dry naturally. But if you are not the patient type, you can build a drying box as will be explained later to accelerate drying. This can reduce drying time down to a day or less depending on the size and the porosity of the mold. If this is still not fast enough, then throw the mold into the kiln on low and cross your fingers.

After air-drying your mold, you may need to heat-cure it and burn out any organic remains of your original model such as wax. To do this, place it in your kiln with model opening up as shown in Figure 22. Slowly heat the mold to drive out any remaining water, not too fast or steam could be produced. Of course slowly is always a relative term. It is relative to how much time you have already put into this mold and how willing you are to take chances with it.

You thermal soak the mold in the kiln first at 225°F to remove physically adsorbed water and then at 350°F to remove chemically bound water. After this, you continue to slowly heat the mold up to about 1200°F to burn out any residual organic materials. This cure may take anywhere from hours to days depending on the size of your mold; hours for molds on the order a few inches, days for ones on the order of feet.

Figure 22. Heat curing two molds in kiln.

You will develop a feel for what's needed with experience. The slower you go, the stronger the mold will remain. This is also a good time to go out to lunch or dinner because final model burnout can produce a lot of noxious fumes. Even well vented kilns and studios may become stinky for a while. These fumes will especially drive off your apprentices so don't count on getting a lot of work done during burnout.

After the mold is burned out, slowly cool it back down to room temperature. If you are able to remove all the modeling material from the mold without having to go into a burn out phase, then you have the option of curing your mold and firing your casting in the same firing cycle. At least air dry it though.

Firing the casting

The bottom of your mold should be flat from the screeding that you did earlier. If it is not flat, the combined weight of the glass and the mold may cause bending which can stress the investment material to failure, i.e. crack it. At this point, you can still file or sand the mold bottom flat if you need to.

This problem can also be alleviated by supporting the mold on the kiln self with a bed of sand. Be sure to use enough sand and to work the mold down into the sand to uniformly support it. If your mold is very large, it should also have some internal support. Check to see that the top of your mold is level as you set it up in the kiln. Otherwise whatever you are making will vary in final thickness. This may not be apparent on small pieces but it certainly will be on larger bas relief ones.

You are now ready to fill your mold with glass frit. If the mold is in the kiln and hot, then this is done as shown in Figure 23. Notice that Dan is wearing heavy clothes and gloves. Because of the heat you will have to go in and get out fast, not only to avoid burns but to prevent thermal shock to the mold. Always turn off the power to the kiln before going into it to avoid shocking experiences.

If the mold is not hot, you have the luxury to set it up on your worktable and take the time fill it by hand in a detailed manner. This process can be as easy as pouring in a single color and size of frit to as complex as using multiple colors and sizes of frit to achieve a variety of effects. For this demonstration piece, we recommend just using the homogeneous frit that you produced earlier when you crushed up that unruly piece of stained glass. If you graded the frit, you may want to place the finer grade first in any accent features.

Figure 23. Filling hot mold with single color frit.

When you fill the mold try to keep the glass to inside the model cavity or mounded over it. You do not want it extending out onto the top surface of the mold. This will result in an increase in glass stringers that you will have to remove later. If the glass flows over the edges and down the sides of the mold, you will surely get cracking that could extend into the model portion of the casting as the glass shrinks down around the mold.

Also whenever filling a mold, try and be careful not to either break pieces off of it or to scratch it as either of these will get investment into the casting. These flaws besides being unsightly can act as crack initiators.

© 2000 James Kervin and Dan Fenton

Basic Kiln Casting Process 21

Once your mold is filled with glass frit and is in the kiln, it is time to cook. A basic firing schedule is sketched out in Figure 24. Note no times are given in this figure. Firing schedules will be discussed in great detail in the chapter on firing procedures.

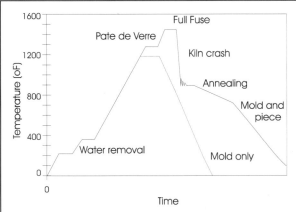

Figure 24. Typical Pâte de Verre firing schedule.

Slowly bring the mold up to the consolidation temperature for your glass. Ramp rates are controlled in order to prevent thermal shocking your mold. If you have not prefired your mold ahead of time, you will have to go slower and dwell at both 225°F and 350°F to remove water bound in the investment matrix as was described earlier. It is during the rise to process temperature that your mold is most likely to crack, because your glass is still fairly rigid and your mold is shrinking some as it reaches the neighborhood of 1200°F.

After reaching the consolidation temperature for your glass, you need to hold the kiln temperature constant until you are sure that your glass has equilibrated and consolidated inside the mold. The mold insulates the glass, causing its temperature to lag behind that of the kiln. You will have to hold the temperature at the control point for about 1 to 2 hours per inch of mold thickness depending on how fast you heated up the kiln.

On the final firing of a piece, assuming that you are doing multiple firings for addition of detailed coloration, you will want to continue on up in temperature to the full fuse point and hold for another hour per inch of mold thickness to get a full density piece. If you are only doing one firing, you would proceed directly on up to the fusing point.

For those visual people out there, you can also peek into the kiln to check on the progress of a firing. There are a couple of key features you can look for:

- The surface texture of the frit. As the glass gets hot, the frit edges soften and lose any angularity. The pebbly texture will gradually change to a smooth pool. It should look like thick maple syrup.
- The color of the glass. As the frit gets to fusing temperature, it along with the whole inside of the kiln will glow bright reddish orange.
- The glass level. As the glass melts, the level in the mold will go down as the voids between the grains of frit get filled in. Remember that we are expecting somewhere between thirty to fifty percent shrinkage in volume. This should be noticeable. You should see the glass vanish down the sprue or go down to the final desired level for an open face mold.
- Flow should be visible from vents. Air vents that we purposefully put into a mold should start to weep glass. This says everything is hot inside.

When you have finished your hold at process temperature, vent your kiln to rapidly crash its temperature down to near the annealing point. Then close your kiln. You will have to hold at this temperature to allow the glass to equilibrate, again for the same 1 to 2 hours per inch of mold thickness. We crash the kiln temperature down at this point to reduce the time spent in the 1000°F to 1400°F range. This is where the glass can devitrify or crystallize on cooling.

You next want to slowly lower the kiln temperature from the annealing point to the strain point for your glass. You will want to do this over a time period of 2 to 12 hours per inch thickness of casting. You will need to hold the temperature constant there for a while to again let the glass to equilibrate with the kiln. This time the hold needs to be only about half as long as before because you have been lowering temperature more slowly. This already helps make the temperature in the casting more uniform.

Lastly, you cool your casting back to room temperature. This can be done much faster than during the high temperature cooling because all the glass is now "hardened" and will no longer build up permanent stress. Just be careful that you do not proceed so fast that you thermal shock it. We will talk more of all this later. Allow the mold to remain undisturbed at room temperature for a couple of hours before you even think of demolding.

It is very important to precisely control kiln temperatures during the firing process. Temperature differences as small as 25 degrees

Figure 25. Carefully break casting from the mold.

can result in significant differences in the final product. Too high a process temperature allows the surface tension of the glass to reduce to the point were it starts to stick to the mold. This can be diagnosed by examining the edges of the final meniscus of the glass in the sprue. If slightly rounded, the temperature was okay; if sharp, the temperature was probably too high.

Demolding and cleaning a casting

Before removing a mold from the kiln, allow it to cool back down to room temperature. If it still feels warm to the touch then you can be sure that it is even warmer on the inside. If it is pretty much encased in investment, it could still be really hot on the inside and you would not know it. So let it completely cool before you start to demold.

When the mold is finally cool to the touch, you are ready to proceed. Since the glass casting will be really sharp, you may want to wear some light garden gloves to avoid cutting yourself. This hazard could be compounded if you embedded wire or wire mesh into the mold to strengthen it.

Although you will probably not raise too much dust while demolding, you may want to consider wearing your respirator or at the very least a dust mask. The silica in the cured investment is just as dangerous as it was before.

To demold your glass casting, slowly break the mold away from the glass as shown in Figure 25. This can sometimes require great patience since you are understandably anxious to see your final product. A light touch is an asset at this point because it is here when breakage is likely to occur if you are heavy handed. Use a small hammer and chisel, if necessary, to slowly break the mold apart.

Try to direct all your blows onto the investment and not the glass. Be especially careful around any thin sections of the casting. That of course assumes that you still remember where the glass is in the mold. Proceed gently since it is sometimes difficult to know exactly where the mold ends and the glass casting begins. To soften your blows, you may want to consider using a plastic or wooden mallet instead of a hammer.

Also be careful as you break away the mold that you do not drop it. The investment can be quite fragile at this stage and with the right blow can crumble apart in your hands. This can cause you to loose control of it if you do not have a good grip on it. Better yet, rest it on your worktable as you break it apart. Block it up if you have to.

If the casting feels hot at all as you break the mold away, stop and let it cool some more. Once the major mass of the investment is broken away, you can pick away at the rest using dental tools. The first sight of the glass will reveal the shape, but be prepared for the surface to look ugly. That's because there is still lots of mold material sticking to the surface. This is especially common with the soft low melting-point glasses or if you cook a casting at too high a temperature.

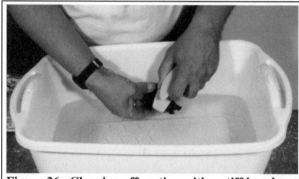
Figure 26. Cleaning off casting with a stiff brush.

After you have the casting free from the mold, don't stick it into water right away to wash off the remaining mold material. Allow it sit and cool for a couple more hours before you clean it. The key to the whole cleaning process is patience.

Once cool, wash the casting in water with a stiff vegetable brush. If you can not get all the mold residue off with a vegetable brush try a small stainless steel bristled brush. Do not use a brass bristled brush on a casting as the brass will rub off and will discolor your piece making it look like a brass sculpture.

Basic Kiln Casting Process 23

Figure 27. Grinding on a casting with Dremel tools.

Once the casting is completely clean, look it over. The first thing that you should notice is that any part of the casting that was in contact with the mold will have a satin finish to it. Those areas open to the air, on the other hand, will be shiny. Also the edges along the sprue or the open face cavity will invariably have some sharp spots where the glass stuck to the investment.

Small imperfections can be removed by simply grinding them away or by sanding with 400 grit wet/dry sandpaper to mimic the natural matte surface texture of the casting. Dremel hand grinders or flex shaft tools with small diamond bits will help greatly with this task. Remember to keep things wet and to wear a respirator with a mist cartridge.

Now that wasn't too bad was it. Did you get bogged down on some of those details. There sure are a lot of them. Don't worry we won't test you, the materials will do that. We will also try to keep the Art police at bay long enough for us to now go back over parts of the process and explain the fundamental details behind the them, so that you will be able to get through it better on your own later.

"Grace & Rapture" 1999
Size: 7½" x 12" x 8"
Pâte de Verre & recast glass
Artist: Philip Crooks
Photo by: Jonathan Wallen

Glass

Almost any glass can be used for kiln casting regardless of particle size. It can be powdered, fritted (granularized), chunked, or even slabbed. You can grind it yourself or buy it prefritted. Figure 28 shows three of the grades of Bullseye frit that is commercially available. As in our example of last chapter, glass used for kiln casting can be broken up soda-lime sheet glass or it can be specially purchased lead crystal cullet.

The type of glass that you choose for your work dictates a number of factors, such as process temperatures, which in turn dictates the investment formulation. If you want color, you can use glass that is precolored or you will have to add color to a clear base glass of your choosing.

If the prospect of mixing colors sounds daunting, then we know which direction you will choose to go. In that case, you will want to search out a line of glass to work with that has as many colors available as possible. As you start mixing different colored glasses, you will also have to be concerned with other factors such as their compatibility. We will discuss many of these decisions and factors in great

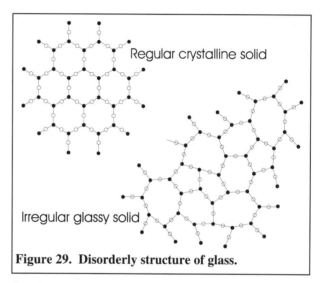

Figure 29. Disorderly structure of glass.

detail in this chapter. There is a lot to learn about glass in order to work with it knowledgeably in kiln casting.

To begin our discussions on glass, let's review what this material that we have all grown to love really is and which two basic types are most appropriate for kiln casting. As part of this discussion, we will examine how a glass's chemical makeup affects the properties of interest to our work and what kinds are commercially available.

Glass Chemistry

Glass is a material made up of a mixture of oxides. The main oxide used in most of the glasses of interest is silica or silicon dioxide (SiO_2). If it is the only oxide present, then the glass is known as fused silica or quartz.

Glasses, rather than being composed of distinct molecules or atoms arranged into an orderly crystalline matrix, consist of atoms arranged in an

Figure 28. Some of the frit sizes commercially available from Bullseye Glass Company.

interconnected, random three-dimensional matrix. Figure 29 illustrates the difference in appearance between these two structures for a two-dimensional matrix of a boron-based glass where each boron atom (black dot) is bonded to three oxygen atoms and each oxygen atom (clear dot) is bonded to two boron atoms. You can see how the crystalline solid looks more rigid. This structure is also more stable and harder to break up once formed.

You can also see from the figure that the structural component for these boron glasses is a flat platelet. The basic structural component for a silica-based glass on the other hand is the silica tetrahedron. This is a four sided pyramid as illustrated in Figure 30 which has four oxygen atoms bonded to each silicon atom, one at each corner of the pyramid.

Together the oxygen atoms act to shield the silicon atom at the center of the tetrahedron from the surrounding environment. Each oxygen atom is part of two tetrahedrons that are randomly orientated with respect to each other. For this reason, the oxygen atoms are commonly referred to as bridging oxygen atoms.

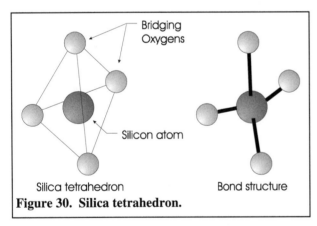

Figure 30. Silica tetrahedron.

The bonds that hold the atoms together in a fused silica glass are a uniform network of strong covalent bonds (electron sharing between atoms) with a lot of crosslinking between chains. This results in a very sharply defined, high-melting point of 3115°F.

This is a much higher temperature than anything you are used to working with in kilnworking. In fact, most of the refractory materials used to line a typical electric kiln will not stand up to repeated exposures to such high temperatures and would self-destruct.

The uniform strong bond structure of fused silica causes it to expand and contract under the influence of temperature much less than most other glasses. The measure of this property is commonly known in engineering as the coefficient of thermal expansion (CTE) and is an important property to our work.

You may be more familiar with one of the terms more commonly used in the glass media — the coefficient of expansion, COE, or the linear expansion coefficient, LEC. We have chosen to use the term, COE, because it is probably the term with which you are most familiar.

To lower the melting temperature of fused silica glass, other oxides are added to break up some of the bridging oxygen bonds by attaching them to a metal atom instead. This terminates one section of the tetrahedron chain making the matrix both less rigid and less viscous. This can happen at random positions in the structure resulting in chains of varying length.

In addition, some of the oxide additives are more electro-positive than silicon which allows them to pull electrons away from the oxygen atoms. This results in forming weaker ionic bonds that lack the directionality associated with the stronger covalent bonds. The metal oxides can also strain the glass matrix if the size of the metal atom is large or small compared to a silicon atom.

When heating up such a modified glass, the weaker ionic bonds and the less-rigid, strained structure allow the glass to become mobile or "melt" at much lower temperatures (about 1800°F). In addition, the variations in chain length of the glass matrix result in different portions of the interwoven tetrahedron chains becoming mobile at different temperatures.

On the macroscopic scale, the effect of these changes to the microstructure is manifested in glass getting soft gradually over a wide temperature range that spreads with increasing amounts of modifiers. Different types of modifiers, as you might guess, will effect this differently. The weaker the modified internal bonds, the lower the softening temperature. Additionally the more modifying oxides that are added to a glass, the shorter the resulting silica chains and the more fluid the molten glass. We refer to glasses that soften at lower temperatures as being "softer" and those that soften at higher temperatures as being "harder" glasses.

You have probably heard or read about glass artists referring to glass as a frozen liquid. What does this mean? Ice is a frozen liquid but it has an orderly crystalline structure and is definitely not a glass. It is referred to as a crystalline solid. What is meant by the frozen liquid analogy is that glass has a "frozen" random structure similar to that seen in the lower half of Figure 29.

This kind of structure is typically associated with liquids, while solids usually have an orderly crystalline structure. The covalent bonds and long cross-linked chain structure of glasses are what causes them to remain disorderly. This structure causes glasses to gradually get more and more viscous (thick and gooey) as you cool them making it hard for them to form the lower-energy crystalline structure.

These crystalline structures can form easier at the surface of the glass because the chains are less restrained there. That is why you generally see crystallinity or devitrification as a surface layer on the glass which can be sandblasted off.

Unlike crystalline materials such as water, there is no single temperature at which glass makes essentially a step change in viscosity as it goes from a solid phase with a very high viscosity to a liquid phase with a very low viscosity. Instead glass has a more gradual change in viscosity with temperature as it goes from a solid to a liquid. An example of this gradual transition is graphed in Figure 31 for a typical soda-lime glass.

The gradual change in viscosity with temperature that you see in the figure is what gives glasses the properties that allow us to work it like we do. If it were like water, instead of slumping, it would puddle. We would not be able to stretch and shape it if it did not soften gradually. Examine the graph of the temperature dependence of viscosity carefully to see the location of the different temperature reference points because we will be referring to each of them when we discuss the high temperature properties of glasses.

Types of Glass

Let's compare the two basic glass types that we are likely to use in kiln casting to fused silica glass. As we do this, we will discuss how their composition affects some of the properties relevant to kiln casting. The range of properties for each of these glasses is summarized in Table 1. Let's start by looking for a second time at quartz glass.

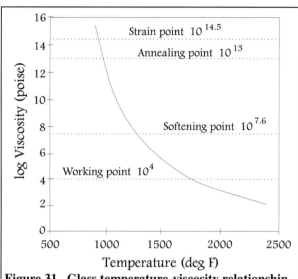

Figure 31. Glass temperature-viscosity relationship.

Quartz glass

Quartz, or fused silica as we have referred to it, is made by heating pure silicon dioxide to about 3137°F. The resulting liquid is so viscous that any gas bubbles trapped between the grains of sand as it melts, come out very slowly if at all. This high viscosity is a result of the rigid three-dimensional nature of the glass matrix.

When cooled, this rigid three-dimensional matrix does not allow it to move around much to seek a minimum volume configuration. This explains its very low coefficient of expansion (COE). The rigid structure also accounts for quartz's relatively high softening, annealing, and strain points listed in the table.

Soda-lime glass

The viscosity of molten fused silica glass can, as was discussed earlier, be decreased by adding fluxes or network modifiers. These are added in the form of metal oxides. The metal oxides that are

Table 1. Range of properties of some basic glass types.

Property	Quartz glass	Soda-lime glass	Lead glass
Softening point (°F)	2876	1280-1350	820-1240
Annealing point (°F)	1983	960-1020	690-980
Strain point (°F)	1753	880-920	650-840
Coefficient of expansion (10^{-7} in/in/°F)	3.1	56-100	47-55
Density (lb/ft^3)	137	153-158	174-338
Refractive index	1.459	1.51-1.52	1.54-1.75

added to make soda-lime glasses are sodium oxide (Na_2O) in the form of soda ash (Na_2CO_3) and calcium oxide (CaO) in the form of lime carbonates ($CaCO_3$). Both of these materials decompose to the oxide by releasing carbon dioxide (CO_2) gas as they react with the molten silica.

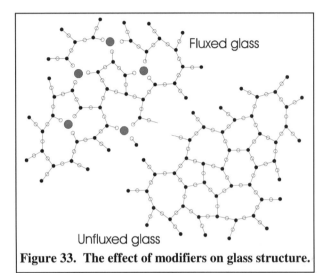

Figure 33. The effect of modifiers on glass structure.

Typical batch compositions for soda-lime glasses are in the range of 70 to 80 weight % silica, 12 to 17 weight % soda and 8 to 12 weight % lime. These materials modify the silica network to shorten and weaken it as illustrated in Figure 33. Varying these percentages will cause variation in properties. As an example, too much lime makes the glass susceptible to devitrification and too little lime results in a glass that is susceptible to chemical attack.

In addition to the soda and lime, other additives are added in small amounts to make the mixture more workable. For example, alumina or aluminum oxide is added to improve the chemical durability. The larger metal atoms of the modifying oxides help to lower viscosity making the glass flow into the molds easier. On the down side, they also result in a higher COE due to asymmetric vibrations of the glass molecules. This requires more careful thermal processing of the finished casting to minimize trapped stress.

The average warm-glass studio artist is probably very familiar with using soda-lime glass from fusing. The brands of glass in common use require processing temperatures in the range of 1575°F to fully fuse. This is on the higher end of the working temperature range of investment materials used for kiln casting. Thus work done with this glass may not have the fine detail desired by some and will commonly have line of flashing caused by mold cracking.

On the up side, there are many varieties of self-compatible soda-lime glass lines (Bullseye, GNA, Spectrum, etc.) that are available with an existing wide palette of colors. Compatible glasses are those with the same coefficient of expansion. Some of these glasses are also compatible between manufacturers, such as Bullseye, and Uroboros. Look for the tested compatible label with Bullseye and Uroboros. Never assume that the glass will fit just by manufacturer alone. All GNA glass has to be tested; likewise with Spectrum.

Besides color selection, there are other reasons that may cause you to want to mix glasses from the different manufacturers — changing the overall viscosity of the glass mixture is one. Adding softer Wasser frit, if you can still find any, to Bullseye frit results in a "softer" frit mixture that flows at a lower temperature. (Wasser flows at a temperature about 100°F lower than Bullseye.) These differences can be important when you are trying to kiln cast without too much flow that might ruin any design you are trying to introduce.

Some glass artists prefer using these common soda-lime glasses since they are trying to produce pâte de verre or frit cast objects that they can fuse together into larger pieces, although this could also be accomplished through the use of glass adhesives without having to worry about compatibility. This technique also allows joining together glass components made from other glassworking techniques such as the piece by Lucartha Kohler seen in Figure 32. This piece entitled "The Myth Maker" incorporates kiln cast, flame worked and sandblasted glass components. Soda-lime glass rather than lead-based glass should be used for projects that might later be used as food bearing vessels because a lead glass may have toxic effects.

Figure 32. Example of a multi-piece work bonded with adhesive.
Artist: Lucartha Kohler
Photo: The artist

Plate glass is one type of soda-lime glass that does not work too well in the kiln casting process because of its tendency to devitrify during the firing process and its subsequent refusal to flow or fuse together. Attempts to try and alleviate this problem by fluxing the glass results in its sticking to the mold. Some glass artists on the other hand, like to use plate glass because of these properties. They like the porous look that they get in pieces made with fine plate glass frit. Another disadvantage of plate glass is that pieces made from it will have a definite green color from the iron impurities present in it.

Lead Glass

Lead glasses are formed by adding lead oxide (PbO) to silica. It is usually used as a network modifier of the silica matrix, but when added in high enough concentration, it can replace silica as the network former. Because of the large size of the lead atom, lead oxide is a good flux that bestows the glass mixture with a larger working range over which the glass changes very little in viscosity. As seen back in Table 1, lead glasses tend to have a much lower softening point than either quartz or soda-lime glass.

One lead glass that may still be available from your local ceramic supplier is Pb 83 lead-based glaze frit from Pemco. Others with essentially the same characteristics are O'Hommel 33 and Ferro 3419-2. They are "colorless" base glasses which are used as ceramic glazes and contain about 60% lead oxide by weigh and can even set off the metal detectors at the airport. This has resulted in more than one embarrassing incident for Dan.

Pb 83 is available as a fine-ground frit for about $5 a pound and fuses into an opaque alabaster-like piece. It has a slight dirty yellow tint to it that is ignored in its intended function. That's assuming you can find a ceramic supply company that will sell it to you with the current phobia of using lead. Most potters are now using lead-free glazes and ceramic suppliers are reluctant to sell lead-based ones to you because they are afraid of its misuse. If you explain what you are doing with it, you may have better luck.

The frit is also available in a number of colors, not all of which may be compatible because they were formulated as glazes to be used in thin layers. It is suggested that, like for any fusing project, you check material compatibility out ahead of time before starting any masterpieces. An alternate option is to prepare your own colors by melting the frit and metal oxide colorants together followed by refritting. This process usually ensures the compatibility of your lead glass frit palette.

Lead glass's main appeal in pâte de verre work lies in its low fusing-temperature, 1040°F for Pb 83. This allows use of fine-grained, low-strength investment materials for development of crisp surface details. Plaster-based investments retain much of their strength at this temperature and therefore are much less susceptible to cracking. The low viscosity of high lead glasses once they are melted requires precise control of your firing schedule since variations as little as 15 to 25°F can result in rapid movement of the glass and loss of your image. Finished pieces made from lead glasses tend to have a matte surface to them possibly from mold interactions or ease of overheating.

Another medium low fusing-temperature (1425°F) alternative is a 24% lead crystal available from Ullmann glass of Germany. This is a true crystal glass without any tint of color. Finished pieces from it tend to have the typical semi-shiny surface of a lead crystal glass.

Some of the common colored glass (Kugler or Zimmerman) used for coloring glass in glassblowing can also be fritted and used for pâte de verre. Since this glass is formulated primarily for glassblowers, most of the colors should be compatible. But again you should probably check compatibility yourself to be sure.

Whenever firing a lead-based glass, you should ensure that you have a oxidative atmosphere (well ventilated kiln) or the lead will precipitate out of the glass and appear as dirty black specks in your finished piece. Your studio should also be well ventilated to get rid of any lead fumes, which may be released from precipitates and are hazardous to your health. Ventilation should be done near the floor in addition to higher up since lead fumes are so much heavier than air.

Also lead glasses and fluxes, as was mentioned earlier, should not be used in any work that might be used as a food-bearing surface. This restriction is to avoid any chances of giving someone lead poisoning. Even though the lead in the glass is vitrified and is in a silicate rather than an oxide form, this does not mean that the lead can not be leached out into foods. Foods that contain acids like acetic acid (vinegar) and citric acid (fruits and juices) can leach out lead. There are also other examples such as coffee, etc.

You may not be aware of it but even some Bullseye glass has a significant enough amount of lead in it that can be leached out. They report that both pink opal (#0301) and salmon pink opal (#0305) leach out lead to levels of 11µg/ml, which is well above the 3µg/ml FDA compliance guidelines for flatware or the more restrictive 0.5µg/ml limit for cups, mugs and pitchers.

Frit

Besides what kind of glass you want to work with, you also have to chose what size frit you want to use based on the desired opacity for your work. You can chose to work with particles as fine as powders to large chunks.

For traditional pâte de verre work, the glass should be finely ground and should contain granules of more than one mesh size to minimize shrinkage during firing. If you were packing perfect spherical particles, ideal packing would be achieved with what metallurgists call a body centered cubic cell structure. The spherical particles, which will pack best, will be of two sizes that have a diameter ratio of 1.4 to 1. This combination, as seen in Figure 34, packs much denser than a single sized frit. If they are of the same glass, then you would use 1.87 times as much of the larger particles by weight as the smaller to get maximum packing.

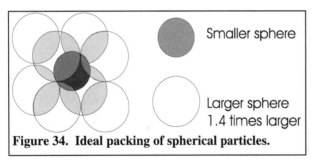

Figure 34. Ideal packing of spherical particles.

Frit is usually measured or sized by passing it through different sized metal meshes, or screens. These screens are characterized by the number of openings or strands of mesh wire per linear inch. This does not tell you exactly how big a particle can go through a given size mesh. But if the wire was of zero thickness, then the particle size passed by a mesh would be the reciprocal of the mesh size. Thus a number 20 mesh would pass particles $1/20^{th}$ of an inch in diameter and smaller. But since the wire does have some thickness to it, a #20 mesh will really pass particles up to about $1/30^{th}$ of an inch in diameter. Table 2 lists example mesh data received from one mesh manufacturer.

Table 2. Mesh sizes from a typical manufacturer.

Mesh size	Wire diameter (in)	% open area	Mesh opening (in)
12	.023	51.8	.0600
14	.020	51.0	.0510
20	.016	46.2	.0340
30	.012	37.1	.0203
40	.010	36.0	.0150
50	.009	30.3	.0110
60	.0075	30.5	.0092
80	.0055	31.4	.0070
100	.0045	30.3	.0055
120	.0037	30.7	.0046
200	.0021	33.6	.0029
325	.0014	30.0	.0017

Mesh sieves can be purchased at your local ceramic supply store but are not cheap. A complete set is typically composed of the following meshes: 4, 6, 8, 12, 16, 20, 30, 40, 50, 70, 100, 140, 200, and 270. Because of their expense as well as the wear and tear that these screens can see from pushing glass frit through them, you may want to consider "calibrating" some common cheap and expendable alternatives. As an example, a spaghetti strainer is about a #10 mesh. You can also use hardware and window screens from your local hardware store. Be creative and save money.

Some glass artists, when using a fine mesh frit (80-100 mesh) will wash the frit in water to get rid of the extremely fine dust sized particles. They then drain the washed frit on a screen that is a grade finer than the last screen used (such as a 120-mesh one in this example). Alternatively you could just screen out the finer particles with a finer mesh and use what remains behind. Those extremely fine particles trap extremely small air bubbles. These small bubbles don't have enough buoyancy to rise out of the melt on their own during the casting process and thus increase the opalescence of the final work.

If you decide to make your own frit, you should be aware that whatever fritting technique you choose to use will produce frit with a predictable size distribution. The size distribution produced by each technique may vary from glass to glass depending on the variations in their physical properties. To get

Glass

near ideal size packing ratios may require using more than one technique and some extra work sizing your frit. But if you are not trying to achieve fine color detail or are not concerned about movement of glass and the resultant blurring of color during firing, then you do not have to be concerned with trying to get dense packing.

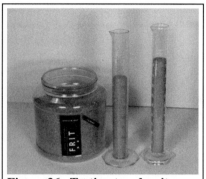
Figure 36. Testing tap density.

To test for how well any mix of frit you make packs prior to filling a mold, you can measure what is called its tap density. As an example, Figure 36 shows the difference between the pour and the tap volume for 101.2 gm of fine Bullseye frit. The pour volume of 96 ml (milliliters or cubic centimeters) is what you get when you pour the frit in. This gives a density of 1.05 gm/cc (weight divided by volume). Then when you gently tap the container on the table to get the frit to settle the volume reduces to 79 ml and increases the density to 1.28 gm/cc. The higher the tap density the better the packing in a mold.

Temperature regimes of glass

To successfully kiln cast glass, you need to understand how its physical properties vary with temperature. This will help you to comprehend why it behaves the way it does, when you have to treat it with kid's gloves, and when you can really push it around. An effective way to think of glass is as a material whose properties vary in different temperature regimes, the ranges of which vary for each type of glass.

The brittle solid temperature regime

You are probably most used to working with glass around room temperature where it is in this temperature regime. Here we see glass as a solid material that breaks in a brittle manner when stressed. Small scratches or cracks propagate through the material with ease.

Glass in this temperature regime expands and contracts in a predictable manner with changes in temperature. The amount of expansion is, as was explained earlier, quantified by a glass's **coefficient of expansion (COE)**. This number which is expressed in units of 10^{-7} inches per inch per degree centigrade, is the average expansion of the glass over a given temperature range. This value is usually determined in the laboratory over the range of 0 to 300°C (32-600°F). The upper end of the brittle solid temperature regime is somewhat higher than this. For soda-lime glasses, it is on the order of 700°F while for lead glasses it is on the order of 600°F.

As the temperature of a piece of glass is changed, the edges of the piece get warmer faster and will try to grow in size relative the cooler glass inside of it. This causes the silica tetrahedron chains to try to slip and stretch past each other. These slippage's build up, straining the natural order of the atoms in the glass, and result in thermally induced stress. In the brittle solid temperature regime, the only way that the stress in the glass can be relieved is by fracture.

The non-brittle solid temperature regime

As the temperature of the glass is increased, the tetrahedral chains become more mobile from increased molecular thermal vibrations and rotations. They start to be able to twist around obstructions that may have pinned them previously in the brittle solid regime. This does not happen all at once but gradually as vibrations of the glass atoms increase with increasing temperature. A measure of this change is the decrease in the glass's viscosity with increasing temperature as seen back in Figure 31.

This increased chain mobility allows any stress that may have built up in the glass to relieve itself. The temperature at which the stresses start to flow in a reasonable amount of time for a given glass is commonly referred to as its **strain point**.

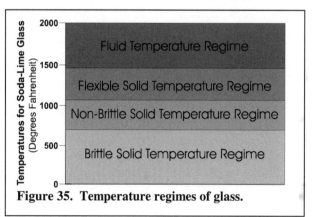
Figure 35. Temperature regimes of glass.

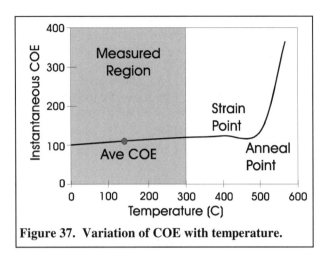

Figure 37. Variation of COE with temperature.

Technically this corresponds to the temperature at which a glass thread has a viscosity of 4×10^{15} poises when measured while cooling at a rate of $4 \pm 1°C/min$. At this viscosity, the stress in a piece of glass takes on the order of hours to relieve itself.

The non-brittle solid temperature regime for soda-lime glasses ranges roughly from about 700 to 1000°F. In this temperature regime, the glass still expands and contracts with temperature change but the COE is different than it was in the brittle solid regime. In fact, a glass's COE is never constant and varies continually throughout all temperature regimes as is illustrated for a typical soda-lime glass in Figure 37.

What you know as the COE is, as we said before, the average value of the COE measured over the shaded region of 0 to 300°C. This value is illustrated as a tiny dot on the curve. As you can also see in the figure, COE increases rapidly at higher temperatures where viscosity is also changing rapidly.

Although the higher mobility of the tetrahedron chains in the non-brittle solid temperature regime allows any build up of stress to relieve itself over time, it is still possible to change the temperature of the glass faster than the stress can be reduced. In other words, you can still thermal shock glass in this temperature regime.

This is also the temperature regime in which glass is annealed. Many artists have been taught to think that annealing (stress relief) takes place at a particular temperature called the annealing point, but that is not quite correct. What that temperature actually represents is a processing compromise. The compromise is that it is a temperature at which the tetrahedron chains are mobile enough to allow relief of stress in a more reasonable time, minutes, without being so high that the bulk glass itself starts to flow in the same time scale.

The **annealing point** technically refers to a temperature at which the viscosity of the glass is 2.5×10^{13} poises measured while cooling at $4 \pm 1°C/min$. At this viscosity, the stresses in your work will be relieved in about 15 minutes. We will discuss annealing in greater detail in a little bit.

The flexible solid temperature regime

As you continue to raise the temperature of a piece of soda-lime glass above 1000°F, it starts to become more and more pliable. If left at this temperature, the glass will droop under the influence of gravity or muscle power. The transition point between this temperature regime and the previous one is commonly referred to as the **softening point**. It varies from glass to glass and refers to the temperature at which a viscosity of 9.6×10^7 poises is reached during the usual prescribed cooling.

The chains of silica tetrahedrons are now so mobile that it is no longer possible to thermal shock the glass. This is the temperature regime we work in when we are slumping or fusing. Glass will start to resemble something more like taffy and its surface also starts to get sticky. It will stick to another piece of glass or to other materials such as separator coatings or kiln wash. These attributes become more pronounced with increasing temperature until about 1400°F and higher when the glass starts to enter the last temperature regime.

The fluid temperature regime

In this temperature regime, hand manipulation of the glass becomes possible. The glass exhibits fluid-like properties and will puddle out if left unconfined. This is the regime in which glassblowers are working when they gather glass from a furnace crucible before they form it into complex shapes.

This is also the temperature regime where full and flat fusing takes place in a kiln. The lower end of the regime is bounded by what is commonly referred to as the working point. It varies from glass to glass and refers to the temperature at which the glass has a viscosity of 9.6×10^4 poises.

Glass compatibility

A glass property near and dear to every hot glass artist's heart is compatibility. Compatible glasses are those that like to work with each other. If you put them together in a piece, they won't fight each other and break everything up as you cool them down. This is the kind of glass that we like to work with because if its happy, we're happy.

If you intend to mix colors and different types of glass, you will have to be very concerned with compatibility because kiln casting two or more incompatible glasses into the same mold is an invitation to disaster. They may be agreeable with each other at high temperatures but will quarrel violently at terrestrial ones where they are not likely to stay together long and their divorce will not be amicable. In order to better understand this disagreement, let's look at the whys and wherefores of glass compatibility.

Coefficient of expansion (COE)

As mentioned earlier, the kinds and amounts of oxide modifiers used in making a particular glass, break up, and strain the silica matrix differently. These changes in turn affect the rate at which the glass shrinks when cooled. This rate, as discussed earlier, is commonly known as the coefficient of expansion (COE) and is usually expressed in the fusing community as a whole number coefficient between 0 and about 120; zero representing the ideal of no expansion and 120 being about the largest value seen.

As an example, the Bullseye compatible glasses all have a COE of about 90. What this means is that in the brittle solid regime, a 1 inch long piece of one of these glasses will shrink on the average 0.0000090 inches (90×10^{-7}) for each $1°C$ reduction in temperature. This is good information to know in case you are ever trying to impress someone at a cocktail party.

Thus a 10 inch piece of this glass will shrink about 0.046 inches in cooling from its annealing temperature (about $950°F$) to room temperature. Similarly a 10-inch piece of Effetre glass with a COE of 104 would shrink 0.051 inches. Therefore, if you were trying to fuse together 10-inch strips of these two types of glass, the difference between the shrinkage of one strip and the other is 0.008 inches. This is about the thickness of 5 sheets of paper. That may not seen like much on our ordinary macroscopic level, but when you compare it to the size scale of reference, atoms, it is enormous. As a result, the piece would literally pull itself apart trying to separate the two types of glass.

Table 3. Coefficient of expansion (COE) of common art glasses.

Glass Manufacturer	COE (10^{-7} in/in/$°C$)
Fused silica (quartz glass)	3
Pyrex, Northstar	32-33
Float glass	81
GNA (Antec 76)	86-88
Kugler, Zimmerman (lead based)	87
Bullseye, Wasser, Uroburos	90
Lenox crystal	94
Spectrum, Uroburos	96
Effetre	104
Pemco Pb83	108
Satake	120

Thus as anyone who has done hot glasswork before knows, you have to make sure that whatever glasses you use in your work expand and contract at nearly the same rate to be compatible. This will avoid breakage upon cooling or sometime later. Some manufacturers cater to hot/warm glass artists and have initiated good quality control measures to ensure a consistent self-compatible glass line.

For small work, small variations (1-3 points) in COE will probable not cause any major problems. But as most good fusing manuals will recommend, you should probably check the compatibility of your glass yourself for large work. This is especially true since the COEs at high temperatures may differ.

Remember the COE of a glass is not really a constant (although nearly so at low temperatures) but instead is an average over the temperature range of 0 to $300°C$. Outside this region, the COE starts to increase rapidly with increasing temperature such that in the annealing temperature range the COE may be as much as 3 times the reported value. If the COEs of the two glass are not matched in this region, this may also cause incompatibility.

Luckily if the two glasses are of similar composition (i.e. both soda-lime glasses or both lead glasses), they will probably match in the higher temperature regions also. Table 3 lists reported COEs for a number of common art glasses.

Determining glass compatibility

The easiest way to be sure that your glass is compatible is to buy glass whose compatibility has been verified by the manufacturer. Many of the expansion 90 and 96 glasses are tested by the factory. Look for the tested compatible label. If you are using frit, they can be a point or two off without encountering any real problems in reasonably sized work. But what should you do if you want to be sure of a glass's compatibility?

The most accurate way of ensuring that two glasses are compatible is to fuse up a test strip. To run a test strip requires a clear compatible base glass against which to test. What you are really doing is testing the compatibility of each of the other two glasses to the base glass. Then by using the Transitive Property of Glass Fusing (i.e. if A is compatible to B and C is compatible to B then A must be compatible to C) we can presume their compatibility.

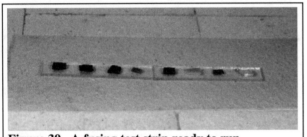

Figure 39. A fusing test strip ready to run.

To run a fusing test strip on sheet glasses, you start by cutting a strip of the clear base glass about 1½ inches wide and about 1½ inches long for each specimen that you are going to test. Thus if you are going to test 6 specimens, the strip should be at least 9 inches long and 1½ inches wide. Next cut ½ by ½ inch square specimens of each of the glasses you want to test for compatibility, although Dan swears that triangular specimens work better to show up incompatibility stress because of their sharper corners. Wash and dry the glass carefully to prevent any contamination from biasing your results. Space the specimens out on the strip of clear base glass ½ inch from the edges and 1 inch from each other.

Keep track of which glass is which since sometimes they change color during firing. It helps to mark both the glass sheet and the test strip by the specimen to ensure no mix-ups. Some marking pens are available (Steel Paint marking pens made by the Alton Company are one version) that have a metallic based ink that will not burn off the glass during the firing and will work well for this process. It can be easily removed after the test if desired.

Next fire the test strip to a flat fuse. This will require a processing temperature somewhere in the range of 1450 to 1650°F for most soda-lime glasses depending on their hardness. After this firing, the thickness of the strip should be almost constant all the way down the strip.

When cooled, examine the test strip for signs of incompatibility by looking for residual stresses in it. To do this, we use polarizing filters like those on expensive sunglasses. You will need two of these filters to conduct the test. These filters have fine parallel crystal assemblages in them that only let light through that is vibrating in one direction, say left-to-right.

If you rotate these filters relative to each other over a light table, you will see that they will get dark and light periodically. They are dark when the filter on the bottom only lets light vibrating left-to-right through and the second filter only lets light vibrating top-to-bottom through. As you rotate the top filter 90° relative to the bottom filter, the crystal assemblages on both are now aligned such that both filters now allow only light that is vibrating left-to-right through. This configuration allows the maximum amount of light through. As you turn the top filter further, they will again get darker until after another 90° rotation the light is again at a minimum.

You are probably wondering how these filters are going to be used to measure residual stress in the glass. To understand how this works, you have to know that light, which is passed through glass that is stressed, is twisted so that the direction of its vibration changes. Light that passes through the rest

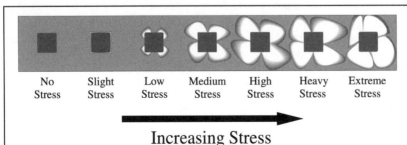

Figure 38. Strip test results with varying amounts of stress.

of the glass that is not under stress is not twisted and thus will be stopped by the filter.

With this in mind, put one filter below the test strip on the light table (or over a light bulb) and the other filter on the top of the glass. Rotate the top filter to the position that lets through the minimum amount of light. Any signs of incompatibility that exist in the test strip will be visible as a halo of light around the test specimens where the light has been twisted by stress in the base glass.

Figure 38 illustrates what this will look like for varying amounts of stress in the glass. The bigger the halo, the greater the amount of stress. The only exception to this rule is for strips that have cracked. In this case, stress will be relieved in the vicinity of a crack. Of course the development of cracks should also give you a clue that something is wrong. Specimens that show only slight stress or less are considered to be compatible with each other.

When doing this test, it is much easier to work with large polarizing filters on a light table rather than those smaller ones on the commercially available "Stressometer." Large filters are available from Edmund Scientific Company and you don't need a letter from the governor to buy them.

A similar test that works well when testing frits or solid sheet glass against a clear control base was suggested by Gil Reynolds and seems to make a lot of sense. It is an edge-fusing test. It gives a larger interface over which the strain can develop leading to higher levels of stress, thus making detection of incompatibility more likely.

Assuming that glasses are $^1/_8$-inch thick, cut two 1 by ½-inch rectangles of each kind of glass to be tested. Place piles of the pieces doubled on a kiln shelf with doubled clear rectangles of your reference glass between each doubled test piece and at the ends of the strip.

If the glass that you want to test is frit, pile it between piles of the clear reference glass. The mass of the piles should be sufficient to allow them to spread out and edge fuse to the adjacent clear stack. The frit pile has to be about 1½ times as high as the base glass pile to accommodate for shrinkage as it consolidates. It may help to have the test strip enclosed in a mold. The interfaces between the clear base and the test glass are then examined for residual stress using polarizing filters as before.

If you do not have any of the clear sheet glass available, you can do the same thing with frit only. Here you definitely need a mold to enclose the glass or you are sure to get a mess. As you fill the mold, use short sections of note cards or paper to help segregate the different frits as you fill the mold. This will help prevent them from mixing as you fill the mold and will result in a more valid test.

Figure 40. Setting up a frit fusing compatibility test.

Try to ensure that each of the frit piles is the same height—$^3/_8$ths to ½" inch high is about right. When finished carefully pull the dividers out. You do not want to induce any mixing of the frit. You may even want to wait until the mold is in the kiln before removing the dividers.

Getting around incompatibility

Sometimes artistically you really want to use two glasses that you know are slightly incompatible together in a piece. Is there anything that you can do to get this to work or are you just courting disaster?

Well there are some things that you can do to minimize the effect of this decision.
- First, always try to minimize the use of any incompatible glass.
- Second, if fusing to full density, you can completely encase the incompatible glass with a fairly thick layer of the compatible glass that makes up the bulk of the casting. This works best if the incompatible glass that you are encasing is of a lower COE so that the interface between the two glasses is under compression after cooling to room temperature.
- Third, try to cast at a lower temperature. Here you will get a sugary like casting where the fused portions of the individual grains are fairly small resulting in lower stress between them because of the small area of contact.

- Lastly, minimize the stress present in the piece is by annealing it very well as we discuss next.

Annealing glass

Annealing is the process of taking a piece of glass to a high enough of a temperature that its molecules are sufficiently mobile to relieve internal residual stress and then slowly cooling the material in a way to minimize build up of any new stress.

In pâte de verre or kiln casting work, we are not usually removing stress from a piece. Instead what we are trying to do is minimize the amount of stress that gets locked into a piece during the cooling process. To learn how to accomplish this, you have to understand how stress gets locked into a piece.

Theory of annealing or stress minimization

As you remember, glass expands when heated and contracts when cooled, so any temperature gradients in the glass result in differences in length of adjacent portions of the glass. These differences in lengths of nearby sections are commonly referred to as strain. They can, in the limit, be thought of as differences in distances between adjacent glass atoms.

Atoms in any material stay together because they are attracted to each other and these attractive forces can be thought of as rubber bands. So as you stretch (strain) the rubber bands, they build up forces (stress) in them that want to push or pull the atoms back to their optimal position. For rubber bands, if you hold them stretched for long periods of time, they start to grow in length and loose much of the force on them. For glass, this can happen at higher temperatures where the attractive forces become weaker but not at lower temperatures.

In cooling the glass back down from the process temperature to room temperature, the glass cools from either one or two outside surfaces as pictured in Figure 41. This sets up temperature gradients in the glass such that the outside is cooler than the inside. The optimum distance between glass atoms increases with temperature as seen by the fact that glass has a positive COE. So the temperature gradient sets up a density gradient which causes the adjacent atoms to yank and pull on each other.

At high temperatures, where the atoms are still relatively mobile, they will flow around to relieve this

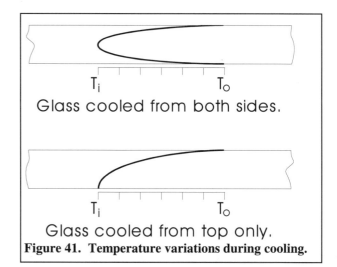

Figure 41. Temperature variations during cooling.

stress just as the rubber band stretches with time. Thus they will move into an area where the rubber bands, or forces, between the atoms are stretched or are in tension. Likewise they will flow out of an area where they have been squeezed together, commonly known as compression. So in the high temperature fluid and flexible regimes, you don't have problems with stress. The glass just flows to relieve it.

At temperatures in the brittle solid regime, which are low relative to the melting point of glass, bonds between the atoms are rigid and can not stretch relative to each other. So when temperature gradients cause stress in this temperature regime by differential heating or cooling, they are stuck with it. But this stress state is temporary and goes away as soon as the temperature gradients go away and all the rubber bands become the same length again. Of course you have to make sure that the stress in the rubber bands never gets so great that they break or in this case causes the glass to crack.

The situation, where you run into trouble and develop permanent stress in your glass is in cooling your glass when the outside is cool enough that the atoms can not move much but the inside is still hot enough that the atoms are free to move. So when the hard outside squeezes on the soft inside, the atoms on the inside will move around to push back evenly on the outside like air pushes on the sides of a balloon. Then as the inside cools a little more, these atoms get locked into position based upon the existing temperature gradient.

As the glass cools and the temperature gradient gets reduced, the atoms in the center of the glass find themselves farther apart than the desired spacing on which the atoms on the outside of the glass are located. Thus they will be in tension, pulling against the surrounding glass with a residual stress, while the outside will be in compression, trying push back against the inside.

If the glass on the outside or the inside is not strong enough, the glass cracks apart. This may happen immediately upon returning to room temperature or, if just barely strong enough, it can occur at a later time when subjected to a stressful environment. This can be in the form of a thermal variation or an insignificant physical impact.

Before you can start annealing, you have to know the annealing temperature range for your glass. As we have discussed, there are three commonly defined points that are important in defining the annealing range — the softening point, the annealing point and the strain point.

The **softening point** marks the transition between the flexible solid temperature regime and the non-brittle solid temperature regime for a given glass. It is the lowest temperature at which a rod or strip of glass will slump over an extended heat soak of a reasonable length of time (an hour.)

The **annealing point** is that temperature at which the atoms are mobile enough to allow all stress to quickly flow out of the glass but not so mobile that the bulk glass will flow.

The **strain point** is the temperature at which the atoms are no longer mobile enough to allow the stress to flow out of a piece in a reasonable length of time. This, as mentioned earlier, serves as the boundary between the brittle solid and the non-brittle solid temperature regimes.

The region in which you have to be careful during cooling is the **annealing region**. It goes from the annealing point down to the strain point. The softening point is important because it is an easily measurable temperature that can, as we shall see, serve as a good signpost for the location of the other two points.

Allowable high temperature cooling rates for cooling through the annealing region are dictated by the chemical composition of the glass, the mass and thickness of the glass, and the mass and type of investment material surrounding the glass. The composition of the glass determines the boundaries of the annealing region for the glass (annealing and strain point) and the ultimate tensile limit that the glass can accommodate. The mass (thickness) and type of investment material used in a casting firing affects the rate at which heat can be transferred into and out of the glass. They define how large of temperature gradients and thus strain will be set up in the glass.

The classical analysis of residual stress and strain resulting from cooling at a given rate through the annealing range was developed for plate glass cooled from both sides by Lillie in 1950. Although not quite correct, because it does not take phase changes of the glass into account, Lillie's equations can be used to estimate the residual stress and strain that will result from cooling a glass of a given thickness and properties down through the annealing range at a chosen cooling rate. This rate dictates the thermal distribution during cooling and thus the resultant stress.

If you want the gory details of this analysis look at his paper listed in the reference section or one of the other technical references such as "Glass Practice" or the "Glass Engineering Handbook." If you want details on the modern theory of annealing, get a lot of No Doze and check out Narayanaswamy's paper.

The formula for stress in layman's terms can be approximated as:

Stress = 0.926 (COE) (thickness^2) (Cooling Rate)

where stress is in pounds per square inch (psi), thickness is in inches and cooling rate is in degrees Fahrenheit per minute. We are not going to try and explain exactly where this equation came from to you with any rigorous mathematics (at least as long as you don't encourage Jim). It is based on assumptions of the shape of the temperature distribution in the glass, the heat transfer rates, the relationship of the high temperature COE to the low temperature COE, and some approximations of the stress/strain relationship.

Solving for cooling rate gives the relationship that:

Cooling Rate = 1.08 (Allowable Stress) / (COE * thickness^2)

In the case where you are dealing with something like a bas relief casting, you are really only cooling from one side and should treat the glass as if it were a sheet of glass of twice the thickness (T=2t). This is the typical case in fusing or in bas relief castings where you are only cooling through the top surface of the glass. Making this substitution into the cooling rate equation results in the following relationship, which is applicable to single sided work:

$$\text{Cooling Rate} = 0.27 \frac{\text{(Allowable Stress)}}{[\text{COE} * \text{thickness}^2]}$$

We apologize to many of you who may have just had this entire explanation fly over your head, but Jim is an engineer after all. For the rest of us, we just shake our head and say "Ya uh huh".

These are the two basic equations that you can use to calculate your own high temperature cooling rates. All that you have to do to apply them is decide what level of residual stresses or strains you are comfortable with and plug in that value along with your glass properties. From this you can develop cooling rate tables for your work.

Plate glass companies usually hold the residual strain left in their glass to below 10 micro-strain (0.00001 inches/inch) to ensure good cutability of their glass. This corresponds to a stress of about 100 lb/in^2 (psi). If we impose similar requirements on strain and use the COE of Bullseye glass. You can calculate appropriate cooling rates that are given in Table 4.

If you want less residual stress or strain by a given factor, cool through the annealing zone by a directly proportional slower rate. Thus in optical glass manufacture, where they want strains on the order of a twentieth of the above 10 micro-strain to avoid optical distortions, you would have to cool twenty times slower. You can probably get away with a little more strain than 10 micro-strain, so if you are in a hurry you might be able to double these given rates without disastrous results.

If you want to develop a table for a different type of glass for which you only know the COE, divide the values in the rate column by the ratio of the new glass's COE to that of Bullseye. Thus for float glass, you would divide each rate by 81/90 or 0.9. (Note: this is the same as multiplying by 1.1.) So the lower the coefficient of thermal expansion is, the faster the allowable high temperature cooling rate will be.

Effect of casting shape on annealing

As we have discussed, this situation is for the general case where we are considering cooling semi-flat pieces from one side. Sometimes we have shapes that are considerably different from this. Are any modifications needed for these cases?

Engineers, like Jim, can calculate the temperature gradients and the resulting stress gradients that are generated in cooling a mold full of glass. All they need is a complete definition of the geometry, the rate of kiln temperature change, the glass and investment properties, and a large enough computer. (They sure enjoy their computers.)

Jim ran a number of calculations to examine the effect of different cooling rates on molds and castings of different shapes and sizes. What he was trying to do was demonstrate how the temperature distributions vary as these conditions change.

In the first portion of this study, he looked at how temperature variations develop in three basic glass configurations: large flat sheets, long cylinders, and spheres. For this study, he looked at how the different shapes affect how fast the center of the glass cools down during an instantaneous crash cooling from 1400°F to 1000°F. To approximate this for the calculation, the outside surface of the piece is assumed to be instantaneously reduced to 1000°F.

Of course, the outside surface does really not instantaneously change temperature like that in real life because. Instead, the outside surface cools by air convection and radiation. This would slow down the cooling process. Thus the actual cooling times would be longer than those estimated here.

Table 4. High temperature cooling rates for Bullseye.

Glass Thickness (inches)	Two Sided Cooling (°F/Min)	Two Sided Cooling (°F/Hr)	Single Sided Cooling (°F/Min)	Single Sided Cooling (°F/Hr)
1/8	76.5	4608	19.2	1152
1/4	19.2	1152	4.8	288
3/8	8.5	512	2.1	128
1/2	4.8	288	1.2	72
5/8	3.1	184	0.88	46
3/4	2.1	128	0.53	32
7/8	1.6	94	0.39	24
1	1.2	72	0.30	18
2	0.3	18	0.08	4.5
3	0.13	8	0.03	2.0
4	0.08	4.5	0.02	1.1
5	0.05	2.9	0.012	0.72
6	0.03	2.0	0.008	0.50
8	0.02	1.13	0.005	0.28

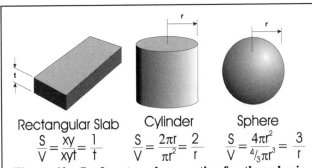

Figure 42. Surface to volume ratios for three basic shapes studied.

As you might expect, each of these shapes cools somewhat differently. This is because the surface to volume ratio increases as you go from a rectangular slab to a long cylinder and then on to a sphere. These ratios, which are given in Figure 42, go from 1/t to 2/r to 3/r. Since cooling occurs from the surface, the higher the surface to volume ratio; the faster the object will cool. Thus a sphere transfers heat out of it three times faster than a rectangular slab and one and a half times faster than a cylinder.

Figure 43 illustrates how these different heat transfer capabilities affect how fast the center of each shape will cool with a step change in temperature of the outside of the shape. It plots the temperature of the center of the glass for each of these three different shapes: a 6" thick sheet (cooled from both sides), a 6" diameter cylinder, and a 6" diameter sphere. In this graph, we see that the temperature variations in large sheets are more than about twice those in spheres and one and a half those in long cylinders.

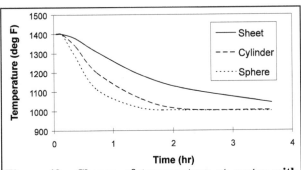

Figure 43. Change of temperature at center with time for different shapes.

Thus, you should be able to cool spheres twice as fast or cylinders one and a half times as fast as sheets to get the same temperature gradient. Since residual stress is a direct result of the temperature gradient in the glass as it cools through the annealing zone, this means that you will then get the same residual stress in each of these cases too.

Effect of mold thickness on annealing

The next question that Jim addressed is what affect does the addition of a mold and its thickness have on cooling of a piece. For this portion of his study, Jim took the simplest possible configuration, that of a spherical glass casting in a spherical mold. To further simplify the problem, he assumed that the mold wall was continuous all around without any reservoir or gating. He then ran calculations on how the temperature variations in the casting developed as it was again instantaneously crash cooled from 1500°F to 1000°F. Here it is the outside of the mold that is instantaneously changed which again is an approximation used for the calculation.

You can see the results of these calculations again for a six-inch diameter glass sphere inside spherical plaster molds of varying thickness in Figure 44. The lowest curve in the figure describes the temperature drop for a spherical casting in a normal half-inch thick mold. The curve shows that it will take the six inch diameter casting about 2 hours to drop to 1100°F as compared to $2/3^{rds}$ of an hour that was seen for a bare sphere in Figure 43.

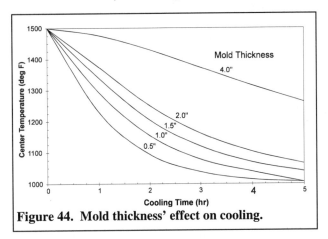

Figure 44. Mold thickness' effect on cooling.

If the mold thickness is increased to 1 inch, the required time increases to $2\ 2/3^{rds}$ hours. Increasing the thickness to 1½ inches increases the required time to $3\ 1/3^{rd}$ hours. Lastly if you used a 2 inch thick mold, you would be required to hold for $4\ 1/8^{th}$ hours to allow the temperature to drop to 1100°F. If for some reason you were to decide to use a 4 inch thick mold, the center of the casting would still be at 1265°F after a 5 hour hold.

The purpose of this example is to illustrate to you the importance of keeping your molds thick enough to have enough strength to contain the glass but not to go overboard. Try to stick near the recommended thickness of ½ inch as much as possible. If you need more strength, reinforce your molds as described in the section on mold reinforcement in the chapter on advanced mold construction techniques. Otherwise you will find that you have to go to extremely long firing cycles.

Determining annealing properties

There are several different ways to determine the annealing point of a glass. The easiest way is to just ask the manufacturer. They usually know their glass better than anyone else since they make the darn stuff. But keep in mind that their interest in annealing is making a product, whether it is blown vessels or sheet glass.

What we are doing in kiln casting typically involves greater localized mass and more thickness variations of glass. This situation requires greater control of the high temperature cooling or annealing cycle than a constant thickness regular shape object does. For this reason their annealing requirements are usually less stringent than ours and they can get away with ramping down through the annealing range faster without having to establish an exact annealing point. We will also want a nice soak at the annealing temperature before proceeding into our annealing cycle to reduce thermal gradients as much as possible.

As we stated earlier, the annealing and strain points of a soda-lime glass can be approximated from the softening point. So if we can determine it, we can estimate the annealing point and the strain point. There is a test that you can run in your own studio kiln to determine the softening point. It can take some time to get accurate results though, sometimes as much a full day with an additional day to test the test.

This test is called a **slump test**. You start by cutting several strips of the test glass ¼" wide and at least 12" long. Place one strip in the kiln with one end sandwiched between two pieces of soft fire brick, so that most of the strip hangs out horizontally over the shelf. Now heat the kiln to about 100°F below the suspected softening temperature and hold it for about an hour to stabilize the temperature distribution in the kiln. Then slowly raise the temperature at a rate of about 50°F/hr.

Occasionally observe the strip through the peek hole to see what is happening. This will disturb the temperature distribution in the kiln much less than lifting the lid. When the test strip appears to show signs of movement, hold the temperature steady to see if it really is slumping. If so, this will be close to the softening point.

To zoom in on the softening point a little more accurately, insert another strip into the kiln and preheat as before. Then slowly heat the kiln to bring the temperature up to 25°F below the previously determined slumping temperature and hold for three hours. If there are any indications of slumping, repeat the test again another 25°F lower. If the test strip doesn't slump, then you know that the softening temperature is somewhere in the middle of the two previous tests. You can make a more accurate measurement if you want by running more tests but that is usually not justified.

The test is easy to run when your starting material is sheet glass but frit requires an extra firing. The frit must first be fused into a sheet that can be cut up into strips.

The annealing point for soda-lime glasses can for all practical purposes be defined to lie about 50°F below the softening point and the strain point is commonly considered to lie about 150°F below the annealing point.

The preceding overview is based on our extensive experience in fusing soda-lime glasses. If you venture into casting of lead glasses, you might find some interesting variations on this. For example, the lab report on Lenox heavy lead crystal indicates a 1161°F softening point, a 826°F annealing point and a 750°F strain point. That's a little bit different than what we just described but lead glasses have a larger working range. Besides what do those guys at the lab know anyway? There just a bunch of scientists and engineers like Jim.

Dan's observations on GNA and he uses a lot of it are that it softens at 1010°F, has its annealing point at 960°F and its strain point at 800°F.

Colorants

If you choose to work with one of the high lead glasses, you may not have the advantage of a ready-made color palette. Instead you most likely will have to make your own colors from clear lead-based frit. Colors in glass are formed by three major processes: dissolved metallic oxides, colloidal suspensions of particles, and inclusion of crystalline materials.

Dissolved metal oxides are dissolved uniformly into molten glass just like sugar in water. They actually become part of the solution and will not settle out again. When the molten glass is cooled and allowed to solidify, the oxides remain dissolved in the solid glass phase. These metallic oxides absorb different wavelengths of light as it passes through the glass. Each metallic oxide absorbs certain characteristic wavelengths of light and passes the rest through.

Colloidal suspensions are a second way that glass is colored. Here tiny particles are dispersed uniformly throughout the melt. Although this may seem the same as the previous method, it is not. This is more like the mix of fine silt in river water. Here the particles are suspended because of the motion of the water molecules, they never really become part of the solution. If you were to take that water and run it through a centrifuge, you would be able to separate them out. In this situation, the size of the suspended particles dictate the wavelengths of light that they can reflect. The remainder of the light is absorbed.

Crystalline inclusions are the third way that glass is colored. In this case, tiny crystalline inclusions of a different material form in the glass when the glass cools. Because they have a different structure they will refract the light differently than the bulk of the glass. This leads to opal-like effects.

Coloring the glass

The method that we will discuss for coloring your glass is not exactly making it from batch as any self-respecting glassblower will be glad to tell you but is closer to batching from cullet. You can color your base glass frit as follows:
- grind it to a fine powder, about 350 mesh
- mix it with a given amount of whatever material you are adding to color it
- pour the mixture into a crucible and remelt it.
- once the melt is very fluid and most of the trapped bubbles have come out, pour it into a bucket of water to refrit it.
- it is now ready to use.

There is a singular joy in realizing that you developed the color in the glass that you are using. We sometimes get off on simple things like that. Of course we've made some good colors and some so bad that we couldn't even blame the dog. The actual color that you end up with after adding a colorant to the glass may vary depending on a number of factors such as:
- composition of the base glass
- concentration of the colorant
- temperature to which you took the melt
- atmosphere of your kiln (reducing or oxidative),
- time at temperature
- how you hold your tongue (just kidding).

Since the main thing that you will be coloring is lead-based glasses, let's review some safety procedures for their use. Lead oxide is a flux added, as you should remember, to make the glass melt at a lower temperature. Since it has been integrated into the silica matrix, it is now a lead silicate, which is relatively inert, compared to the original oxide. But we still advise caution in handling it, especially as a fine powder, because it can be easily inhaled or ingested and once into the body, lead may be leached out of the powder just as it can be by some foods.

So wear a respirator whenever dealing with these fine powders, especially when ladling them into a hot crucible. Not only can some of the lead be released as fumes but also some of the coloring oxides that you will use. Many times these are more toxic than the lead. Vent your kiln and your studio, or at the very least use a fan and an open window.

For the lead glaze bases that we have discussed — Pemco Pb 83, O'Hommel 33 and Ferro 3419-2 — you will only have to heat them to about 1600°F to get good dispersion of the colorants within the glass. At this temperature, they can be poured like honey on a hot day.

Table 5 lists some color recipes that we have found work fairly well for us. The exact amount of colorant to add may take a little experimentation. You may be surprised how little is actually needed to color your glass. The amounts that we give are for coloring 2 cups of glaze base which will just about fill the average crucible to the top.

The colorant amounts are given as ½ gram measures but are actually added as medium size palette knife measures. Hey what the heck, we are usually in a hurry and we cook the same way too. Besides a little uncertainty usually won't hurt. If you come up with a good recipe, you may want to dust off the old triple beam balance and make more accurate measurements later.

Table 5. **Metal oxides commonly used as colorants.**

Color	Shade	Colorant	Measures
white	opal	vanadium pentoxide	12-15
yellow	pale to mid	silver nitrate	1-3
	yellow-green	potassium dichromate	2-3
	champagne	cerium oxide titanium dioxide	60 20
purple	pale	manganese oxide or manganese carbonate	2-3
	grayed	manganese carbonate & copper carbonate	10-12 4-6
blue	light turquoise	copper carbonate	2-3
	med turquoise	copper carbonate	6-8
	deep turquoise	copper carbonate	10-12
	pale	cobalt oxide	¼
	medium	cobalt oxide	½
	deep navy	cobalt oxide	1-2
green	yellowish	iron chromate	5
	jade	vanadium pentoxide	6-8
	dark green	chrome oxide	2-3
	dark grass green	chrome oxide	6-8
	rich green	cerium oxide & vanadium	10-12 2-3
black	purple	manganese carbonate	12-16
amber	yellow	potassium dichromate	6
	medium	iron oxide red	6-10
brown	dark	red copper oxide & nickel carbonate	2-3 2-3
	dark amber brown	lithium carbonate & nickel carbonate	30-40 4-6
gray	pale	nickel carbonate	4-6
	medium	nickel carbonate & manganese carbonate	8-10 4-6
	gray-blue	potassium nitrate & cobalt carbonate	20-30 2-3

Modeling

The model is the original object whose shape you want to cast in glass. It can be made from a number of different materials. Since many glass artists do not have an extensive art background and may not be familiar with common modeling materials, we decided to expand upon this subject in further detail in this chapter.

The materials most commonly used for modeling in direct casting of single-part waste molds are wax, clay, plaster, and organic materials such as small vegetables or, for the more gruesome of you, small animals. All but clay and plaster can be burned directly out of the mold. If you make multi-part molds, then almost any object can be used to form an impression in a reusable mold as long as there are not undercuts that will trap the model in the mold. Let's look at each of these modeling materials and discuss how they are used.

Modeling With Wax

Wax has been used for modeling effigies for thousands of years. Wax models of religious figures have been found in tombs of ancient Egyptians. Wax has many properties that make it the ideal material for modeling. It is durable, lightweight, clean working, and does not shrink, crack, or dry out. The media can be worked in a number of ways: sawing, carving, casting, handworking, etc. Wax has the advantage of allowing development of intricate details even down to fingerprints. That's one way to prove you did the work.

The process of making an original out of wax, casting a refractory mold around it, and then melting the wax out of the mold is the basis of **cire perdue**, or lost wax casting. This technique is used by craftsmen in a number of trades (bronze sculpture, jewelry, glass, etc.). It is especially beneficial when making multiples of an object.

Working directly in wax can sometimes involve a considerable amount of time depending upon how detailed you try to get and quite a few tools depending upon how much of a tool collector you are. With wax, you don't have worry about having undercuts or draft in your model because you are just going to melt the bulk of it out and then burn out any residue that might remain trapped in the mold.

You will want to learn a little about the chemistry of wax in order to understand which additives are added to the wax to soften it and make it more flexible. Additives are also added to change other properties such as melting temperature and plastic range. These additives include such things as animal fats, oils, tallows, gums, resins, and fillers.

Types of waxes

The term wax is loosely applied to describe many different oily solids and there are many different kinds of waxes. They vary in a number of properties; the most important of which are slipperiness, plasticity, melting temperature, and hardness.

Waxes are often classified according to how they have been prepared, as being either natural, modified, or compounded.

- **Natural waxes** are those that can be found directly in nature from animal, vegetable or mineral sources.
- **Modified waxes** are natural waxes, which have been chemically or thermally altered in some way.
- **Compounded waxes** are a mix of a number of different waxes to get the properties that you want.

The choice of which wax to use in your work depends upon which process that you are using.

Most glass artists work with microcrystalline wax. This is also the wax of choice for most bronze sculptors, because it is cheap and easily available. For finer detail some of the harder jewelers' waxes, of which there are two major manufacturers — Kerr and Ferris, may work better. Even beeswax and paraffin can be used for modeling.

It is all right to mix waxes, you will just get a new compounded wax formulation, with slightly different properties. Most waxes have already been blended to give them specific properties such as toughness, flexibility, viscosity, and melting point. Slight changes should not hurt them. Just take care that you keep them clean and try not to overheat them when blending.

Besides being available in different formulations, waxes are also available in different shapes. These include sheets, blocks, rods, wires, bars, tubes, etc. Feel free to use whatever shape or formulation of wax that works for you. The Art Police will not even cite you if you decide to construct different portions of the model out of different types of waxes.

If you're thrifty like us, you can try to reuse as much of your wax as possible but you will invariably end up with small investment fragments in your wax that can make modeling and mold cleaning more difficult over time. For this reason, many glass artists choose not to reuse any of their wax and advise others against doing so. After all, wax is probably one of the cheapest materials that we are dealing with in the whole kiln casting process. So trying hard to save money there may not make sense if it results in losing a piece of work.

Microcrystalline Wax

Microcrystalline wax is a relatively recent natural wax with properties similar to natural beeswax. It was originally an unwanted petroleum by-product. In order to keep motor oils fluid at low temperatures, the relatively high molecular weight materials had to be removed. The material removed was a hard waxy material that at first no one had any idea what to do with. Now-a-days, the production quantities of this unwanted material far exceeds that of any of the other natural waxes. It is now the standby, as was previously mentioned for those craftsmen that do bronze casting in a real foundry situation, even if they only cast baby shoes. It has good strength, a fairly high melting point and is relatively flexible—all properties that are important in our work.

Three types of microcrystalline wax are available from our local arts supplier (Douglas and Sturgess) under the name of "**victory wax**". Heck if we know who won what, but it saves harvesting honeycombs. There are three types and they differ mainly in hardness and melting temperature.
- **White** is the hardest of the three.
- **Yellow** is slightly softer and also works well as an easel wax for stained glass. (You do remember stained glass, don't you?)
- **Dark brown** is the softest of the three and is the one preferred both by foundry personnel and sculptors. This is the reason that it is the type of microcrystalline wax that can usually be found in the larger art supply stores.

Microcrystalline wax is typically sold in 1 or 10 lb. blocks and is cheaper than the jewelers waxes because of the larger volume of wax produced helps bring down its price. If you need wax in bulk, you can also order microcrystalline wax directly from the petroleum manufacturers. If you do so, you will have to be prepared to specify it by complex technical jargon that we do not intend on presenting here in detail.

Good sculptural microcrystalline wax will have a melting point in the range of 140° to 180°F. In addition, it will be fairly soft. Hardness of a wax is typically determined by measuring how many tenths of a millimeter a "standard" needle that has been warmed to a given test temperature (usually 77°F) will penetrate into the wax — the higher the number, the softer the wax. For sculptural work, you want a wax with a needle penetration in the range of 20 to 40 tenths of a millimeter.

Red Wax

Another sculptural wax that is a little harder than microcrystalline wax is red wax. It is also sometimes referred to as "French" wax. Our first guess was that this means it goes well with wine and can only be used a little at a time but what do we know. This type of wax is commonly used in sculpture by building it up over a soft wire backbone and appears to be particularly suited for casting. Red wax is more expensive than microcrystalline wax and because of its hardness needs to be worked with sharp or heated tools. It is similar to the carving waxes, which we discuss next.

Modeling

Carving Waxes

Carving waxes are really hard waxes, which are able to hold sharp details. They are hard on the order of a hardwood. Unlike a hardwood, they do not have any grain to them but can be worked with many of the same tools. They cut cleanly with a knife or a saw, as well as file like a metal. Their melting temperatures are so high that they can even be worked with powered hand tools. The harder they are though, the more brittle they are. You have to be careful not to inadvertently split them during carving. They are available in blocks, tubes, rods and sheets of various thicknesses.

One manufacturer is Ferris, which makes three different carving waxes for jewelry work that have been dyed different colors each with slightly different properties. The three types are:

- **Green** is the hardest of the three. It allows development of greatest detail, but is also the most brittle. This brittleness usually limits its use to small objects. It melts at 220°F, and immediately becomes very fluid.
- **Purple** is more flexible and a little softer. This makes carving of it a little easier. It melts at 230°F and first becomes viscous before becoming fluid. Unlike green wax though, once melted, it becomes too soft to hold fine detail.
- **Blue** is the most flexible and carvable of the three. It melts at 230°F and remains very viscous. This allows control when adding wax to a modeling project as will be discussed later.

It should be obvious but worth mention that the colors of waxes by other manufacturers will not necessary have the same meaning.

Mold-a-wax

Mold-a-wax is another wax type produced by **Ferris**. This wax is generally used by cutting it into thin sheets or slabs. These sheets are then softened in warm water and pressed into molds or shaped around objects. This allows construction of thin walled objects or vessels. The wax is made in two types. The first, **black**, is the harder and higher melting temperature of the two. The other, **red**, is very soft.

The problem with using this wax is its softness. It can inadvertently become distorted as you as you remove it from whatever you formed it on. This problem can be minimized by prior application of a release agent on the object you are modeling but even with this precaution it can still be a problem. It is hard to put further detail in a model of this wax because it is so soft, and it is hard to add extra wax to a mold-a-wax model because it melts at so low a temperature.

Casting Wax

Casting or injection wax is a hard wax, which melts at a relatively low temperature to form a low viscosity fluid, and also sets up quickly to its original hardness. It is used primarily for casting replicas from a master mold as will be discussed later and is often introduced into the mold by injection.

Set-up Wax

Set-up wax is primarily used for wax build-up. It is a low melting temperature wax that is used by melting and the dripping it onto your work. It is too soft for filing or carving and melts at too low a temperature to use motor tools on it. It is primarily worked with hot tools, although it can be carved with a sharp knife while still warm.

Ferris makes two types of setup wax, blue and Perfect. Of the two, **blue** is the harder and most workable with carving tools which makes you wonder why they call the other **Perfect**. A hard yellow microcrystalline wax also works well for this purpose.

Sticky Wax

Sticky wax is a soft, gummy wax that melts at fairly low temperatures. Its main purpose is to act like an adhesive in bonding pieces of wax together. You use it to bond individually constructed components together into a single model.

To use it, simply warm one of the objects that you want to bond in one hand and a little sticky wax on the end of a tool in the other. Spread the sticky wax onto the piece. Then rewarm the sticky wax region of this piece and the other piece that you want to bond to it. When ready, press the pieces together. This should hold them together well enough to cast a mold around it. Sticky wax is very similar to the wax that is used in stained glass work to hold pieces of glass to a glass easel or to a lamp mold.

Water Soluble Waxes

Water-soluble waxes are used in situations where you may want to create an object with a hollow portion. It is brittle and cannot be carved or filed easily, but it can be built-up like a set-up wax. Other waxes can be added around it. When your object is finished, you just place the whole thing in warm water and the water-soluble wax will dissolve away. This is especially useful in making things like beads where you want a hollow in the model to make it lighter. You will probably never use this wax though because even for beads it is easier and cheaper to insert a hole former of something like fiber paper into your finished mold.

Paraffin

Paraffin, another petroleum byproduct, is the hard white wax that is commonly found in your supermarket. Besides being used for sealing the tops of jelly jars (that in itself is a use of which we highly approve, especially if we are on your mailing list), you will use this hard wax primarily as an additive in compounding other waxes to harden them or to lower their melting temperature.

Paraffin is so hard because it has been highly refined to contain only the higher molecular weight material. If you want, you can try to model with it directly. Try working it as if it were a carving wax. Be warned that it tends to be even more brittle than most carving waxes, so work carefully. It can also be used as a cheap substitute for a casting wax because of its hardness and low melting temperature.

Beeswax

Lastly, if you want that back-to-nature feeling, you can work with the real thing, beeswax. It is a nice soft wax available in a number of different flavors if you so desire. This can make melting it the first couple of times a fragrant proposition that everyone will enjoy. You can get it from a number of sources but price may vary considerably. It was twice as expensive as microcrystalline wax at one art supply store visited during preparation of this work.

Plastecine

One wax-like material that is not compatible with wax is plastecine. This material is actually more like a clay. It should not be mixed or compounded with your other waxes if you plan on trying to recover and reuse them because it will not really mix with waxes. Instead of getting a smooth uniform mixture as you do when mixing other waxes, you will end up with a crumbly material sort of like pie dough that is completely unusable for anything and will have to be discarded. (Unfortunately our pie crusts are not usually much better.) If you do not reuse your wax, as is recommended, you are not bound by any such restrictions.

Modifying waxes

You may want to modify your wax to try to make it more suitable for whatever process with which you have chosen to work it. You can modify your waxes in a number of ways.

- To soften them, try adding squishy oily materials like petroleum jelly or cocoa butter.
- To harden them, try adding harder waxes.
- To give them more body, try adding sawdust. For sculptural wax applications, clay or whiting are traditionally added but these will not burn out. So we do not recommend their use.
- To color them, try adding oil-soluble dyes like wood stains.
- To lower their melting point or viscosity, try adding a low melting temperature wax like paraffin.

Working with wax

Wax can be worked in lots of different ways. We'll go over some of them but we guarantee that you will come up with other ways to abuse this material. Try to choose a technique that is appropriate for the wax with which you choose to use. Wax is fun to work with and if nothing else you can always use it to make candles.

Melting Wax

Wax can be melted and cast into shapes to be used in modeling. Typical shapes include sheets, rods, blocks, bowls, etc. There are wax melting pots available from jewelers' supply houses but they tend to be rather expensive. An old pot on a hot plate will work in a pinch by setting it on low, but must be watched so as not to overheat the wax and possibly start a fire. This problem is further aggravated if you use a burner with an open flame like a gas stove or a propane camping stove to heat the wax. It is hard to tell by eye what the temperature of the wax is because wax at 500°F looks exactly like just melted wax at 150°F.

Overheating a wax can change its properties. It might break down the chemical structure of the wax. Any volatile components may vaporize from the melt. Such changes may make a wax more brittle.

Modeling 47

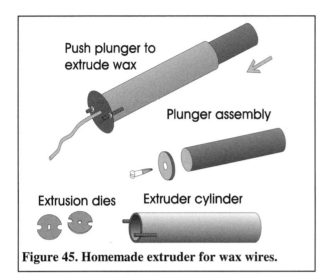

Figure 45. Homemade extruder for wax wires.

Forgetting your wax on the stove can be definitely bad for your business, unless you have a very understanding insurance agent. If you melt wax directly on the burner, there is also the possibility of it reaching its flash point and exploding into flames. If you should start a fire in your wax pot, the best thing that you can do is turn off the burner and smother it. For a fire restricted to the pot, you can smother it by putting on the lid. On a slightly larger fire, you might have to use a fire blanket. If you knock over the pot, you may have to go at it with a fire extinguisher.

To avoid overheating your wax, use a double boiler arrangement when heating it on a burner. This prevents overheating the wax to the point where it starts to vaporize and create an explosive hazard. Remember to add more water to the bottom boiler pan as it boils away otherwise you can really ruin that pot. This is one of Jim's standard cooking tricks.

One economical way to safely melt wax is to use a cheap crock-pot. It will take a while for the wax to melt, but you can depend on it not to boil the wax even if you leave it on all day. You can usually get a good used crock-pot for about $5.

Never leave hot wax unattended, especially if any small children have access to the studio. Even if not a fire hazard, it is still a serous burn hazard. When melting wax, make sure that it is dry prior to being added to the melting pot. Water is heavier than wax and will immediately sink to the bottom of the pot. There it will flash boil and splash hot wax all over the place. Unfortunately this is also probably the one time that the watched pot actually boils and you end up with a face full of hot wax.

Even dry wax can be hazardous when reheated because it expands as it melts. On a burner, the bottom portion melts first and builds up pressure. Then when the top layer gets thin enough, hot wax may erupt through it. Whenever melting a full pot of wax, it is suggested that you make a hole through the wax to the bottom of the pot. This will allow any pressure that might build up as the wax melts to bleed off. Another way to prevent problems would be to tip your pot when cooling so that the liquid hits the bottom of the pot. Then when you melt it next time you are melting part of the top surface right from the beginning and it can not build up pressure.

Because of the hazardous nature molten of wax, we suggest that it may be a good idea to keep a cover on your wax pot to prevent accidents. A clear lid is preferred so you won't always be taking it off to see what's happening. You might also want to consider wearing safety glasses. Hot wax will stick to your skin and continue to burn you until you can cool it off. So keep some cool water around to cool off any body parts that may get splashed with hot wax.

Also don't spill any hot wax on your carpet. Wax exploding in your face may be preferable to your spouse's or landlord's response to wax stains in the carpet. Residual coatings of wax can be removed from tools and pots with mineral spirits but it is hard to remove from carpet. The only way that we know to do that is to first dig the bulk of it out with a dull butter knife or popcycle stick. The rest can sometimes be wicked out by placing a plain paper bag over the spot and running a medium iron over the bag. Good luck if this happens to you. But having a pot of hot wax around is so handy for many things that you may want to risk it.

Casting Wax Shapes

To cast wax into **sheets**, pour hot wax from your pot out onto a slab of plaster saturated with water by soaking it overnight. There should be no standing water in the mold or it can cause it can flash into steam and spray wax if you pour the wax too hot. Wax will tend to splatter anyway if it is poured too hot so make sure that you cool it some.

The plaster mold should have ridges around its edge to contain the wax. Make sure that the walls have enough draft on them to easily release the wax sheet. Balsa wood makes a good modeling material to cast molds for this purpose. It is easy to shape and smooth.

You can also cast sheets of wax on marble or glass. Just remember to use a little release agent and make a clay dam to contain the wax. Also if you melt wax out of a mold into a pan of water, it forms sheets of wax that could be reused. These sheets of wax can be warmed and shaped. They are especially useful in developing cloth-like textured objects.

All sorts of **wires** can be made from soft waxes by using an extruder such as is used for cake decorating. Construction of such a tool is described by Tim McCright in his excellent book on casting metals. The concept is illustrated in Figure 45.

This tool can be constructed by using a section of pipe about an inch in diameter and about a foot long to form the barrel of the extruder. The plunger is made from a tight fitting piece of doweling with a tighter fitting rubber washer screwed to its end to prevent wax from leaking out the back end as you depress the plunger. Brass dies of 18 gage or greater can be mounted on the end of the barrel using bolts which have had their heads cut off and have been soldered to the sides of the pipe with their ends extending past the end of the barrel. The shape of the hole in the die will dictate the cross sectional shape of the wire.

The tool is used by pouring molten wax into the barrel and forcing it out through the die with a slow uniform pressure into a dish of cool water. If the wax in the extruder hardens up on you, heat the whole extruder's worth back up in a pot of boiling water. When you use this tool, you should wear heavy gloves to protect you hands from burns. You should also wear splash goggles, and a plastic face mask to protect your eyes and face in case hot wax spurts out of some orifice unexpectedly.

Rods of wax can also be made by pouring wax into soda straws and cutting the straws away after the wax is cool. Or you can make some more plaster molds with grooves in them for this purpose.

For casting more **complex shapes**, you can make quick molds out of clay. The rough clay shapes can be carved to include detail into your wax. To cast wax in the mold, you pour in some molten wax from your double boiler and roll it around the mold until it completely covers the inside.

To cast a hollow shape, some of the extra wax that has not solidified in a mold can be poured out after it was allowed to cool for a couple minutes. This process produces hollow castings as done with slip casting in ceramics. You can continue to build up the thickness of your wax original by repeated applications of wax until you get the desired thickness. You can use this same technique to cast wax into those pottery slip molds if they are well coated with a release agent prior to introducing the wax. Otherwise the wax will stick to the mold.

For developing good control in casting wax, it helps to get a feeling for your wax's **plastic range**. This is the temperature range from the liquidus, the point at which the solid wax will melt, to the solidus, the point at which a molten pool of wax will start to skin over and harden. The larger this range, the easier it will be to work with your wax. This range can be measured using a candy thermometer.

If you don't have a thermometer, you can perform a simple test with a paintbrush to check to see if the wax is cool enough to pour. Just dip the brush into the wax. If it hisses and burns, then the wax is too hot.

When pouring wax, do not let it get to be too hot. Keep it in the plastic range. Don't have any standing water in your mold. Pour the wax slowly and evenly into the mold. As it cools, the wax will change color slightly. This will occur around the edges of the mold first and it should start to release at the same time. If it doesn't, then you probably forgot to prepare the mold with release agent ahead of time.

Carving wax techniques

Hard carving waxes are worked by various cutting techniques. These include sanding, filing, scraping, routing, carving, and sawing.

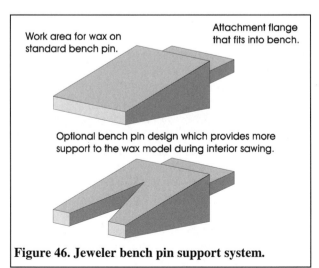

Figure 46. Jeweler bench pin support system.

Always try to support your work as much as possible when working on it to prevent cracking the wax. This can be accomplished by using a V shaped bench pin support setup like jewelers do. Two versions of these bench pegs are illustrated in Figure 46. The peg in the back of the picture fits into a hole in your workbench. This frees your hands and allows better access to all sides of the wax piece as you carve, saw or file it.

Saws

Carving waxes can be sawed roughly to shape using a coping saw. The friction from the saw blade sliding through the wax actually melts the wax as it cuts. If an ordinary coping blade is allowed to stop for any length of time, the wax can harden and trap the blade.

The most versatile saw blade for use in waxwork is a spiral wax blade which has been manufactured such that all its teeth project outward from the blade in a spiral. It cuts a wide enough curf through the wax that it does not lock up in the wax. But unfortunately, it also wanders some as it cuts and therefore is only good for roughing.

For sharp corners, you might like to use a #4 piercing blade. They are used because coping saws and hacksaws can not get into tight corners. For removing interior areas, you can drill the wax to form a starting hole, thread the blade through the hole and connect it to your saw frame.

Carving

Sharp details can be carved using an X-Acto knife. Be careful when carving the wax or you may fracture it. Take only small short cuts. By holding the sharp edge of the blade perpendicular to the wax and dragging the knife across the surface of the wax, you can shave small amounts off the surface in the form of fine curls of wax. This leaves a nice smooth finish to the surface and allows a lot of control as you work. Unless of course you are the heavy-handed type like Jim and always nicking yourself in the morning when shaving.

When carving wax, try to keep the wax flexible or you may snap it. As you cut, make sure that your blade is sharp and new as they dull quickly, especially if you are using a plate glass pallet. When you make a cut, be aware that many times you may leave a wax burr on the side of the cut. This can be removed by cutting parallel to the back

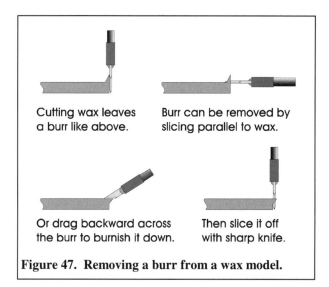

Figure 47. Removing a burr from a wax model.

surface or by burnishing it to the side with a backward scrap of your blade before cutting the burr off the side as illustrated in Figure 47.

Cut away from corners to avoid fractures. Do not try to cut all the way through the piece of wax in one pass as this is sure to cause fracture of the piece. Instead, make a number of shallower cuts. This method is also much easier to control. Do not hurry too much or the blade may wander. But at the same time try not to work too slow because too much handling at one time will result in softening of and leaving finger prints in the wax. Softened wax can be carved with a sharp X-Acto knife after it has been softened by dipping the knife in hot water.

Dental tools are also commonly used to shape wax and are handy for scraping, carving and scratching the wax just like they are used on teeth. You can even pretend to be a deranged dentist as you work. "Oh, you wanted that tooth! That's too bad!" Dental tools can also be used for melting detailed designs or applying wax.

Never buy your dental tools new, they are too expensive. Instead ask your oral hygienist to save you old ones that are ready to be retired from regular service. They don't have to be sharp to work well on wax. Dental tools can also be found at flea markets but still tend to be overpriced.

Wood carving or linoleum carving tools also work well on soft or hot wax. If you need to hollow out areas, this can be done using miniature gouges like those used in woodcarving or turning. Don't heat any of these tools in a flame though if you are ever

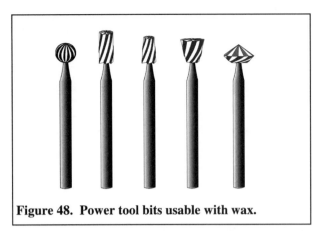

Figure 48. Power tool bits usable with wax.

plan on using them again for their original purpose as this will cause them to lose their temper.

Motor driven tools

Motor driven tools and burrs can be used on hard carving waxes. The burrs should be coarse-toothed such as those illustrated in Figure 48. Special three-bladed burrs shaped like small propellers have also been designed for use on wax and are available from your friendly neighborhood jewelry supplier. Even these burrs can heat up quickly and clog.

Melting and clogging can be minimized by working motor driven tools in short strokes where you lift the burr off the work between strokes. You can also just repeatedly touch the same spot lightly allowing the burr and wax to cool between touches. If you can control the speed of the shaft, slow it down.

Filing wax

Once the carving wax model is cut roughly into shape, you can then use one of the many different wax files that are available to remove saw marks and finish shaping it.. Some of these are illustrated in Figure 49. Coarse-tooth files are best because they cut quickly and are resistant to clogging. For those hard to reach places, you can get a set of really small wax working files from a jewelry supply house. For final smoothing, switch back to a fine-tooth metalworking file. These will cut slower but will leave a smoother final surface. For smoothing fine details in your final polish, we suggest using a set of rifling files.

To make you wax easier to file or carve, it is suggested that softer wax pieces be made harder by sticking them in the freezer for a while before you start working on them and cooling them as necessary.

Fine toothed files will have to be cleaned often using a file cleaner to prevent clogging. This is a flat wood brush with short metal bristles. Clogging of your files can be minimized by applying a light coating of a powder like talc, cornstarch or whiting to the file. A light spray of silicone lubricant or mold release also works well. Whatever you do, do not try to melt your wax off. This will only make cleaning them a lot harder.

Sanding

To get a final polished finish on a wax model, you may want to touch it up with a little bit of fine sandpaper. Be careful though because it can leave grit in the wax. A final buff can also be done using a soft cloth or nylon stocking. The open mesh acts as the sanding instrument. Again you may have to cool that wax as was recommended for filing.

Adding Wax

Even though you can cut almost any shape from a single block of wax, sometimes that is not the best use of your time, much less the expense of a large block of wax. Many times it makes more sense to cut a number of small blocks of wax and stick them together to rough out the shape of your piece. They get stuck together using heat. Allow the wax to completely harden after joining before you start carving it. This will be apparent from its returning to its original color and will take a couple of minutes. Remember, joining can be facilitated by using a low melting point wax like sticky wax to help join the pieces together.

To apply the heat for bonding or adding wax, small

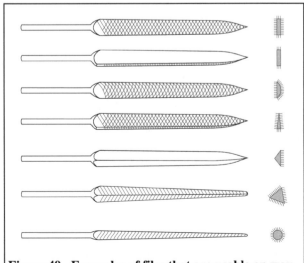

Figure 49. Examples of files that are usable on wax.

Modeling 51

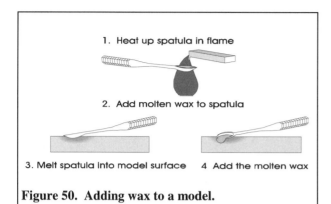

Figure 50. Adding wax to a model.

spatulas heated in an open flame are usually used. The simplest flame source is one of those short stubby candles available from the supermarket. Most artists prefer to use an alcohol lamp like those available from jewelers' supply houses. They burn denatured alcohol, available as shellac thinner from the hardware store, and provide a much cleaner flame than candles.

You bond small delicate wax pieces together using these spatulas to apply the heat to the joint in a controlled manner. You first tack the joint at a number of points by touching it with the hot spatula. To permanently bond them together, you take the hot spatula and repeatedly bury it about $^{2}/_{3}{}^{rds}$ of the way into the joint. Continue to work your way along the joint, reheating the spatula as necessary.

After joining, add extra wax as necessary to fill any depressions in the joint by heating some wax in the bowl of the spatula and again burying the full spatula into the crack. When you pull it out, it should leave most of the wax in the bowl behind. Use this procedure to completely fill in cracks or to build up areas with extra wax. This technique is necessary because just like flameworking glass, hot wax will not bond to cold wax. When done on one side of the joint, flip the wax over and do the same to the other side.

Cheap soldering irons make good hot wax sculpting tools and can be temperature controlled by hooking them up to a plug with a dimmer switch. For attaching larger pieces of wax, try heating up the surfaces to be joined together in a flame and then pushing the two molten surfaces together. Touch up of the joint can then be done with a spatula as described above. Be careful with the molten wax though, because as we said before, it can stick to your skin and burn you.

Handworking Wax

Handworking of sculptural waxes is done when they are warm and pliable. Start by getting a flat surface to cool the molten wax on like a marble or plaster slab. Apply oil to the slab to act as a release agent. Ladle as much molten wax as you want to work with out of your wax pot and onto the slab. Allow it to cool slightly, but do not let it harden. Lift and flip the sheet on the slab trying not to crack it. Cool the other side slightly.

Now with oiled hands, knead the wax into a malleable ball and work it until it has the right texture, that of soft candy. Model the wax into your basic shape. Use your hands and tools to develop this shape. Add more soft wax as necessary using your fingers and pressure. More wax can be added with a hot spatula as described before in areas where pressure is not appropriate. Heated small loop-ended clay tools can be used to shape the wax by melting it and causing it to flow. They can also cut really soft wax just like clay.

An alternate method to warm wax for handworking is to stick it in an old coffee can and warm it with a reflector-type drop light, like those used for photography. Use the drop light like a lid for the can with the bulb on the inside. The reflector should be slightly larger than the can so it will not fall into it. If the reflector has a spring clip on it, this can serve as a good handle to lift the light.

Figure 51. Alcohol lamp used to flame polish wax.

By a little trial and error, you may find that you will be able to control the softness of the wax by the light bulb wattage that you use and the length of time that you apply it. This will vary with the amount of wax you are heating and the desired working temperature. You can tell when microcrystalline wax has been warmed sufficiently because it will change from the dark chocolate color to a milk chocolate one. It won't taste any sweeter though.

© 2000 James Kervin and Dan Fenton

There are a number of ways that sculpture wax can be worked. It can be rolled out between your hands like clay to form elongated shapes. It can be pinched between the fingers to form flat details. It can be twisted and stretched to elongate it. It can be worked with pointed objects like pencils. It can be smoothed with your fingers. It can be bend or folded. It can be sliced with knives, gouges or carving tools. (It might help to cool the wax before trying to carve it.) What you can do to it is only limited by your imagination.

Flame polishing

After you are done modeling, you may want to flame polish the surface of your wax piece by passing it over a flame. You can do this with an alcohol lamp. The cheap ones that look like Austrian crystal with multiple surfaces as shown in Figure 51 work well because you can tilt them and work on both sides of the flame. There are also alcohol lamps with little nozzles where you can blow through a hose to direct the flame.

If you want to really get fancy and have about $50 burning a hole in your pocket, you can get an adjustable micro-torch called a Blazer Piezo from a jewelery supply house. This torch has a fast and focused little flame, so do not go too slow when fire polishing the wax or you may have a professional overrun. It is fueled by the same butane cans used for filling cigarette lighters.

Texturing a wax surface

Instead of flame polishing the surface of your wax model, you may want to texture it. This can be done by heating up the wax to about 300°F where it gets uniformly shiny and then pressing whatever has the texture or design that you are trying to imprint on the wax surface into the wax.

You could use leather-working dies to put in letters or decorations. You can also press found objects or rough textures from nature like rocks into it. Avoid any object with undercuts or you will find yourself breaking off wax as you remove the object. Also be careful not to burn yourself. We suggest that you wear some heavy gloves when trying this.

Modeling With Clay

Let's look at our second modeling material — clay. Modeling with clay is less complicated than working with wax. There is no meltdown or burnout, no hot caldrons of volatile material and no fumes. Clay is cheap and can be recycled, as long as it is not allowed to dry and it is easy to keep mold chips out of it. There are many artists who prefer the feel of clay and find wax difficult to control.

The disadvantage to clay is that there are limitations to the shape or detail that can be put into it and cast as part of a single part mold. Because it does not melt like wax, it has to be dug out of a mold. This is an important consideration and access must be provided to all clay. Jim has literally spent hours digging clay out of molds with tight deep spots in learning this point. Figure 52 was one of those problematic casting made from a clay model. For this reason, we use clay mainly as a modeling material for open face castings.

Figure 52. Casting made in difficult to clean mold.

What is clay

As James Mitchner would probably explain to us in excruciating detail, clay is a natural decomposition product formed by the weathering of granite and igneous rock. This decomposition is primarily brought about by the action of water over years and years. Water, as rain, abrades the rock just as sand does glass in sandblasting. In changing from water to ice during winter, water that has seeped into cracks in rocks will expand and further crack the rock into smaller and smaller pieces. As ice or glaciers, water grinds the rocks upon each other to form fine gravel.

Plants also assist in this process by sending roots down into cracks, which expand as they grow to help break the rocks apart. Some minerals in the rock are soluble in water and once exposed end up being leached out by it, thus weakening the rock matrix.

Finally when the rocks become small enough, they get ground up further by the tumbling action of streams and surf. During transportation by streams, the individual grains of rock are also sorted by size. As the water slows down, it drops bigger grains first while carrying the smaller grains further downstream.

Since the weathering process happens everywhere on earth, clay is quite common. Its composition is primarily determined by the original rock composition and this composition really does not vary all that much from place to place because the composition of the earth really does not vary all that much. It is about 58% silicon and 15% aluminum, in the form of silica (SiO_2) and alumina (Al_2O_3).

The silica and alumina are chemically combined with water in the ratio of 2:1:2. Note this chemically combined water is not the water, which gives clay its plasticity. If you allowed the clay to get bone dry, the chemically combined water will still be present.

Clay also contains oxides of other materials such as iron, magnesium, calcium, sodium etc. in smaller amounts. These other materials add to the color and physical properties of the clay.

The purity of a clay depends on whether it was formed in place or whether it was formed during transportation from one site to another. Clays that have been formed in place are called **primary** or **residual clays**. The primary mechanism for this formation is through chemical action of water on the rock, most commonly on feldspar.

Because primary clays have not received the added benefit of mechanical action and sorting that takes place during water transport, they are composed of a mixture of grain sizes both large and small. Also because they have not been mixed with other materials during transport, they tend to be whiter and purer than other clays. Both of these factors make them coarser in texture, more difficult to work, and gives them higher refractivity. For this reason, as we will see in the next chapter, one common residual clay (Kaolin clay) is often added to investment mixes to make them more thermally resistant.

The other clays, those that have been formed and transported by the forces of nature, are referred to as **secondary** or **sedimentary clays**. They are more common than primary clays. But because they have been transported and mixed with materials from many locations, they have a more complex composition than a primary clay.

Transportation of the clay allows picking up inclusions of impurities like iron oxide which give many clays their reddish color. It also leads to inclusions of organic materials, which increase the clay's plasticity. As mentioned earlier, the transportation also results in segregation of the size of the grains of sedimentary clays making them finer and more pliable. All of these differences affect the properties of a clay.

Desirable clay properties

The primary property of interest for modeling is a clay's **plasticity**. This refers to the ease with which a clay can be shaped and how well it retains its shape. Plasticity is believed to be a function of the distribution of size and shape of the individual grains within a clay. These grains are thin plate-like structures surrounded by an even thinner layer of water. The layer of water is held in place by the attractive adhesive forces of the large surface area of the grains. Clays with a higher content of fine grains are more plastic because the grains slide over each other easier.

A clay's organic content also adds to its plasticity by inclusion of many smaller softer particles made by bacterial action on the organic content. These act as lubricants on the clay particles. The reddish brown clay that reminds one of dirty baby diapers is a good one for modeling.

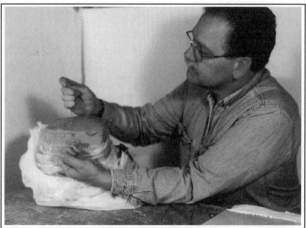

Figure 53. Wire tool used to cut slabs from clay block.

Most of the clays that you will find are actually a mixture of a number of different types of clays and have been formulated to get the optimum properties for one of a number of ceramic process. Such mixtures of clays are referred to as clay bodies and are formulated to control factors such as color, plasticity, shrinkage, firing temperature and surface properties.

The type of clays you should use for modeling are **throwing** or **modeling clays**. These will have good plasticity and have been formulated to minimize the cracking that occurs during drying. Stay away from stoneware and porcelain clays because they are less plastic and coarser. If you want to know more about clay see "Clay and Glazes for the Potter" by Daniel Rhodes.

We don't think we need to tell you where to buy your clay. We hope! While you are there you might want to buy some of those wire tools cut slabs off a big block of clay. They are inexpensive and work great. They are used as shown in Figure 53. And don't forget some of the loop tools that we will be taking about in the next section. You will need them for working those fine details that are too small for fingers.

Working With Clay

Working with clay is very easy. It is no harder than making the mud pies that we probably all did as children. (Did yours taste as bad as ours?) Any type of clay can be used, but the water-based products rather than the oil-based plasticine types are usually preferred. This is because the plasticine type of clay is harder to clean out of the mold and any residue can cause carbonization on the surface of the glass.

In working with a water-based clay, you can make it softer and more plastic by adding additional water to it. When storing it away at the end of the day, encase it in plastic bags or sealable boxes to prevent it from drying out. If the clay does start to dry out, it will lose some of the thin layer of water that gives it its plasticity. This will make it hard to work. You really want to keep it moist.

If your clay does dry out, it can usually be reconstituted by adding a little water and working the water in. As clay dries it also shrinks. This can cause cracking between sections of varying cross sectional thickness or from shrinkage down upon some internal support.

You can help keep your clay moist while you are working it by wetting it with your fingers or by spraying a mist of water on it. An old perfume bottle, plant spray bottle or an air brush works well for this task. If you have to leave a clay model on which you are working out for a while, cover it with a plastic bag to help prevent it from drying out. If it dries out, it will be hard to work. In addition, any new clay that you try to add to the dried model will not want to stick to the dried clay.

Coiling

Clay can be worked in a number of ways, many of which we all learned as children. One of the easiest is with coils. Make coils from the clay by rolling it between your hands or on a flat surface. These serpentine coils can then be formed into other shapes like pots.

In kindergarten, you probably made many such masterpieces by starting with a small circle and successfully building up layers of damp clay coils of larger diameter circles until you had a bowl. You may have even been one of those advanced students that then added coils of successively smaller diameters to form a pot.

If you were even more advanced, you probably then smoothed out the coil texture to a nice smooth surface. But if you were like Jim, you probably had big gaps between your coils and they probably weren't circular either. So what, that's art too.

Pinching

Another way to form clay into shape is to pinch or press it together with your fingers. As you squeeze the clay, the grains will flow and come into intimate contact with each other to form a uniform mass. This causes it to stick together. Bending or folding can be used to further shape it. After you stop pushing, it retains its shape. To help prevent the

Figure 54. Rolling clay slabs out to uniform thickness.

Modeling 55

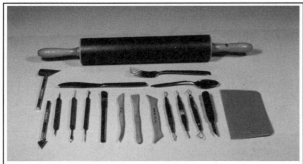
Figure 55. Various clay modeling tools.

clay from cracking as you pinch it thinner, it helps to keep your fingers damp.

Slabbing

When working to make flat clay shapes of multiple layers, the best technique is that of slabbing. Flat slabs of clay can be formed by flattening it out with a rolling pin. Uniform thickness can be achieved by using guides of constant thickness under both ends of the rolling pin as illustrated in Figure 54. To prevent the clay from sticking to the surface on which you are rolling it out, you can dust the surface ahead of time with something like finely-ground fired plaster or you can roll it out on plastic wrap which can be removed later.

The slabs are cut into two-dimensional shapes and overlaid upon one another to form three-dimensional shapes. As you lay one slab layer onto the next, moisten the contact surfaces to allow them to stick better. If you are going to stand these shapes on edge, allow them to stiffen a little before cutting them out.

To join slabs, roughen the surfaces that you are going to join and bond them with a layer of clay that has been mixed with so much water that the mixture is almost creamy. Smooth out any of the bonding cream that seeps out of the joint with your finger tips.

Pressing

In making clay vessel models, you may sometimes want to use a bowl for an aid which you cover with clay and shape as desired. When you do this there are two things that you need to do. The first is make sure that your mold will not trap the vessel because of a small opening. Second apply a coating of oil to the vessel so that it can be easily removed from the clay after casting the mold.

If you forget this, the vessel or any other object can still be worked free from the mold by getting the clay very wet and gently rocking the object back and forth. You may have to continue adding water to the clay as it will seep away though the walls of the mold. This will leave some pretty mushy clay, which may be hard to clean out of the mold afterward.

Handworking

The most versatile tools of any clay modeler are his or her fingers. They can do almost anything. But there are details that are too fine to put into a model with fingers. Those finer details can be carved into the clay using knives and some of those inexpensive wooden modeling tools that are illustrated in Figure 55. You can also use common objects like popcycle sticks, pins and nails to carve details. To smooth out areas of a model, use flat spatulas or rounded spoon-like objects. Wire loop tools are good for carving out bits of clay. They come in different shapes and sizes.

Figure 56. Large cast sculpture composed of bonded pieces. Artist: Mark Abildgaard.

If you are making a larger sculptural model, you might want to use an armature to support the clay. You can build one from heavy copper wire or wood. Anchor it to your worktable. Make sure that you plan out ahead of time how you will release the armature from the mold.

Mark Abildgaard likes to make fairly large kiln cast sculptures such as the one he is standing next to in Figure 56. He forms them by bonding several kiln cast components together using Hxtal epoxy. To ensure that the final cast pieces will fit well together, he works with fine textured sculptural clay and lets the components dry to a leather hard state before casting the investment molds. This allows him to play with the fit of the parts and ensure that it is true.

After investment, any residual clay has to be cleaned out of the refractory

© 2000 James Kervin and Dan Fenton

mold or it can stain the final glass casting. This process is made easier if a little separator compound is added to the outside of the clay model. Remove the clay as soon as possible after the investment sets while it is still moist. You know when the investment is set because it is hard and no longer warm to the touch.

If the clay is allowed to dry in the mold, you may never get it out. Remember clay can not be melted and burned out like wax. It has to be carved and washed out. Start removing it by using the loop tools to dig it out in chunks. Then wash out whatever residual clay remains on the plaster mold. This can be done using cool running water and a soft brush like a tooth or paintbrush.

Preparing a clay model for investment

After you have pretty much finished your model, you might want to work to modify its surface texture. To most artists, this means smoothing it up. Do this by working it over with dampened tools or fingers to remove ridges and fill in creases. Follow this up with brushing it down with a soft wet brush for a nice slick finish.

Other artists instead may want to roughen up the surface, maybe by jabbing at it repeatedly. Just remember that if you do decide to do this, the rough texture will tend to trap investment, which will be hard to clean off of the casting.

At this point you would be well advised to let the clay dry out some so that the surface of the model will be harder. If the surface is too wet when you invest it, the clay can get into the pores of the investment. It will be hard to clean out and will invariably result in some staining of the casting. But at the same time, you want the clay to be soft enough so that you can dig it out easily. Leathery is about the right way to describe how you want your clay.

Even when you wait for the clay to get leathery before investing, you can still get some penetration into the investment. What can you do the minimize this. You could seal the surface of the clay with a waterproof barrier.

Traditionally this is done by painting it with a light coat of shellac thinned way down with alcohol (shellac thinner). These are mixed in a one-to-one ratio. This mix spreads evenly and does not build up thick enough in crevices to remove detail. But if you are not quick enough in applying it, the first section will curdle when you come around to finish the coating. This is also true if you try to add a second coat of shellac over wet damp clay. We would recommend that you spray a clay model with a couple of coats of acrylic paint instead. This will dry harder and the second coat does not dissolve the first as with shellac. Either of these options will leave a thin impermeable barrier to water but we find spray acrylic easier to work with.

These are not release layers. For that we will usually apply something like green soap over the coating just before investing.

Modeling with plaster

Plaster is a modeling material that has been used to make models since the time of the Egyptians. But despite this glorious past, plaster has fallen into disuse as a modeling material. It is still a good material for making simple forms, especially linear and axi-symmetric ones.

Since we will talk in great detail about plasters in the chapter on investment materials, we will not talk about what they are here. Instead, we will discuss a couple of different ways of making plaster models. These include sculpting, turning, and forming.

Sculpting plaster

When sculpting plaster it usually necessary to build up your model on an armature of some sort because the plaster does not have much strength in its buildup form. How big and complicated an armature depends upon how large of a model you are trying to make. For small pieces, a wire armature is adequate. For larger pieces, you may want to build a form out of wood or polystyrene. Anchor these down to a woodwork surface to handle the weight of the plaster. It helps to shellac this surface so that it will not steal water from the plaster.

Figure 57. Plaster modeling tools.

So bend the armature into the general shape that you want for your model. Pad it out some with wadded newspaper that you tape to the armature. Put a first layer over this of plaster bandage to give it some strength. Now mix up some Plaster of Paris as we discuss for plaster-based investments in the last chapter. In modeling we want to extend the working time of the plaster so you will probably want to add some wallpaper sizing to slow down its setting. We would recommend about one-half ounce of sizing to every pint of water. It will also last longer if you let it slake for about ten minutes and use it with out mixing.

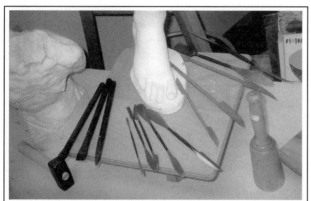

Figure 58. Plaster sculpting tools.

Dampen the plaster bandage or any previously applied plaster. Pick up some thickened plaster with your hands or a tool like a trowel or putty knife. Spread it onto the model in thin layers of about a quarter of an inch or so more on horizontal surfaces. If you try to apply it too thick, it may slump on you. Try to organize your work as a continuous build up of layers, allowing just enough time between layers to allow the plaster to partially set. Rewet built up plaster as necessary. Do not try to smooth the layers until you are roughly to size because the layers will stick to each other better. If necessary add some plaster bandage for strength. Try to have any of this far enough from the final surface that you will not carve into it.

Carving plaster

After you have your model roughly to shape it is time to smooth it and carve any details into it. Carving is easier to do when the model is damp. This helps prevent fracture of the plaster. Carving of plaster is done with knives and chisels or tools that look a lot like some of the shaping tools that we used for clay. Shape flat sections with chisels, pushing by hand toward the center of the piece to avoid breaking off edges. Do not drive the tools into the plaster too deep though or you still might fracture it. Shape curved sections by digging in with your knife and then smoothing with a claw chisel.

When you have your shape roughed, set the model aside to let it dry some before starting to refine it. We do this because damp plaster will clog the finishing tools. These include rasps, surform tools, saws, drills, or sandpaper. We usually do as much shaping as we can with rasps and surform tools first as they cut into the plaster best. If you did not let the plaster dry enough and the tools clog up, use a wire brush on them to clear off the plaster.

For final smoothing work first with a finer rasp. Then switch to coarse sandpaper and finally to fine sandpaper. Again clogged tools or sandpaper can be cleared with a brush. It does not necessarily have to be a wire brush this time though.

Turning axi-symmetric plaster shapes

Axi-symmetric shapes are ones who have circular cross sections. Thus the distance from their axis to the outside edge is constant at each cross section. This allows them to be turned. Turning is a form of carving that is done on a lathe. Basically you spin the work piece about its axis, either vertically or horizontally and carve the shape by resting a tool against. You may have seen bowls or table legs turned on a wood lathe. You can do something similar with plaster.

Figure 59. A plaster turning box.

The difference between how plaster shapes are turned and how wood is turned is that plaster is so soft that you can cut the whole contour of the shape at once. To do this, you make a template of the shape that you are forming. This template is half of a piece's cross section as show in Figure 59. They are cut from a sheet metal like zinc-plated steel that you can get from the hardware store. You cut the contour with a hacksaw.

The template is mounted to the turning box on which the plaster piece it turned. A turning box is basically a box without a bottom or a top. Two sides of the box have a slot in them through which the shaft is fitted. One slot has a stopping plate at one end to keep the shaft in position. Both of the slots may also have retaining clips to hold the shaft in place. One end of the shaft has a crank mounted to it to crank it. The template side of the box is notched so that the template is mounted at the same level as the shaft. It is convenient if the box is made large enough to mount over a baking pan to catch any plaster that comes off of the shaft.

So now we are ready. Start by wrapping the shaft in some plaster bandage or coiling on some string. This will give the plaster something to cling to. Mix up a batch of plaster to about the same consistency as we described for sculpture. Apply some of the plaster to the shaft by pouring it over while turning the shaft. Apply it slowly up and down the length of the shaft.

Figure 60. Using a jack to form a plaster model.

As enough plaster builds up on the shaft, the template will start to scrape against it and the desired shape will start to emerge. The plaster that builds up on the template should be scraped off frequently. The easiest way to do this is sometimes to remove the template off the box and scrape the plaster back into the bowl. If you run out of plaster, scrape out what you have in the pan under the box. If you still need more, mix up a second batch and apply it to the first.

After a while, you will have a complete piece and the template can be removed. You may want to continue to rotate the shaft occasionally until you are sure that the plaster is completely set up. Then, if you want, you could turn sharp details into the piece using a scraper or smooth it out using wet-and-dry sandpaper. Continue turning the shaft during either of these processes.

Then let the piece completely set up, remove the shaft from the box, and twist the piece off of the shaft. Plug up the holes on both ends with a little newspaper and plaster. Let them harden up and sand them down.

Forming linear plaster shapes

Linear plaster shapes are ones that have the same cross sectional shape down their whole length. One portion of this shape will be a flat base. Like turning this technique uses a template to cut the surface. The difference is that this time the template is dragged down the length of the piece instead of having the piece rotated against it. The template may form the whole surface at once or just part of it.

This template is mounted to a wood brace and together they are called a **jack**. The jack has a rail on it that is used to guide the jack down the piece. The jack has to be sturdy to ensure the trueness of the form as you scrape it down the length of the piece. The plaster model is build up on a base plate that has its basic shape.

If the piece is a continuous long piece, then the base plate will be a long rectangle and the jack will be pulled along the length. If it is to be octagonal, then the base plate will be an octagon and the jack will be dragged down each side. The base plate could even be a circle in which case the piece will be axi-symmetric again with the axis pointing up.

The base plate should be sealed with shellac and a release agent to make removal of the final plaster model easier. If you want a hollow model, you could build up an initial core from clay that you can dig out after the plaster is formed. Use the jack as a guide in building up this core so that the model will have close to equal thickness walls. Try not to make the walls too thin though.

So now mix up some plaster as before and apply it to the base plate. Scrape the jack down the length of the piece by running the guide down the length of the base plate. Scrape any plaster off of the jack and depending upon the shape of the model move on to the next side. Repeat this until you have formed the model to your satisfaction. Add extra plaster if necessary.

Allow the model to set up and then smooth it out with wet-and-dry sandpaper.

Readying a plaster model for investment

Plaster models should be sealed before casting an investment mold around them. For this we use a couple of coats of thinned down shellac. This works much better on plaster than it did on clay.

We always have a batch of shellac that we have mixed up ahead of time and stored in a tightly sealed container. When mixing the two stir them do not shake. The last thing we want to do is create a bunch of small bubbles in the mixture that will collect on the model. Remove any air bubbles by blowing on it or using a sharp tool. Do not apply it so heavy that it runs or drips.

When applying the sealer coat, it helps to position yourself so that you can see the reflection off the wet shellac as you apply it onto the model. This will help you to ensure that you apply a nice even and complete coat. This is easier to see on the first coat. If the pallet is of a porous material coat any area of it that will come in contact with the investment also. If do not, you may find release of the mold from the pallet to be difficult.

When you apply the shellac, use soft brushes of high enough quality that they will not leave hairs on the model. Soak this brush in paint thinner when not in use. This will to keep it soft and pliable. Squeeze out any thinner before using the brush.

After the shellac has completely set up, it also helps to coat it with a release agent. Any of the usual release agents will work fine.

Models From The Real Thing — Organics

Don't worry. You don't have to go to the health food store for this option, although we all could probably benefit from frequent visits there. Organics in this context means anything that was once a living organism and can be burned out of a mold. This eliminates things like seashells or coral, which are primarily calcium deposits left behind by once living organisms. If you really want to capture their shape,

Figure 61. Collection of Pate de Verre peppers.

they can be used to construct master molds from which you can then cast wax replicas.

Of the organics that can be completely burned out, vegetable matter seems to be the most popular one used. (We guess because most of us don't like handling dead animals or insects.) Of these, the vegetation to which we relate best is food. In other words, this means vegetables or fruits. Many vegetables have very interesting shapes and textures. We are sure that you have noticed this during your many years of playing with your food and now pâte de verre gives you the perfect excuse to further partake in these mealtime games.

Go to your supermarket and browse through the fresh vegetable section with a new eye. Some vegetables, like brussel sprouts, have boring shapes and are better left for eating. (Dan says that he actually likes them.) Look carefully, pickling cucumbers have a nice texture that can prove interesting. Asparagus always works as does cauliflower. Watch out for things like broccoli or artichokes though because cleaning out the fine details in the investment mold after burnout is difficult. Figure 61 shows a collection of Pâte de Verre peppers that Jim put together.

Be wary of using woods or nuts. They can be used as models but they can smoke up your studio if you attempt to burn them out of a mold. Burn a walnut out in your studio and your neighbors will have the fire department down on your case in no time flat. If you must do so, be sure to vent your kiln area.

Woods also swell when exposed to water and could thus crack your mold. This can be prevented by coating a wood model with a sealer such as a spray on lacquer. Two light coats will usually be sufficient.

Keeping these comments in mind, carved woods can produce some interesting patterns. It is best to use the softer woods, like pine, balsa or cedar, because they burn out more easily. Fir has too much pitch and like oak is not all that easy to carve.

The other thing to be aware of with organic models is that during burnout some them can be quite fragrant. Most will just steam but there are a few, like garlic, that beside driving away the vampires will also drive everyone else out of the studio in a matter of minutes. Curiously enough some of the ones that you would think might be problematic, like hot peppers, have no impact at all.

Anyway, it is always good practice to vent the kiln and the kiln area, even if you have to open all the windows and use fans. Otherwise apprentices and employees will consider switching to something like a career in advertising. The worst part about bad burnout fumes is that they always seems to happen right at lunchtime, leaving your stomach a little unsettled.

You might want to get out into your garden to search for inspirational specimens. Some suggestions might be an unopened rose or other flower buds. Look for those oddly shaped vegetables or fruit. A favorite is peppers that have grown with shapes like faces, complete with noses and mouths. Dress them up a little with some wax eyes and glasses and they make great novelty pieces.

Also think about what might fit into local events. How about making small pumpkins with faces carved into them around Holloween or corn on the cob at Thanksgiving. Maybe you have some sort of local festival that will provide inspiration. In our area we have garlic, pumpkin, artichoke, and asparagus festivals. We also have festivities like the erotic-exotic ball but those subjects may be better left alone.

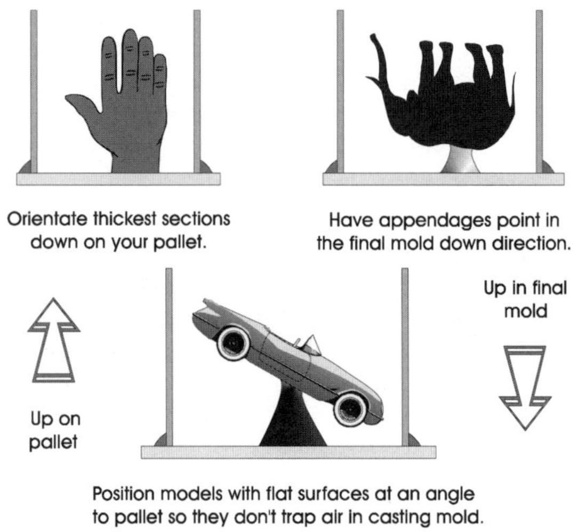

On a walk in the woods, you can find an abundance of possible specimens. Young pines cones, small twigs, or tree bark pieces are just a few of the many candidates. Small animals like banana slugs, the unofficial California state mollusk, are also fair game. Tree leaves are usually too thin to work with effectively, but they can be dipped in wax to give them more mass. Unfortunately this will result in loss of detail, so break out the wax carving tools to carve them back into the model.

A trip to the seashore or lake may uncover other great finds like driftwood, fish, seaweed, or shells. The great part about any natural thing is that the viewer can easily relate to it. For a really eye catching piece, you might try juxtipositioning natural objects. How about a fish swimming through pine cones, or birds flying through coral and seaweed.

The advantage of replica casting is that you can be creative by selection. Just because you can not draw (stained glass) or carve (clay, wax, or wood) doesn't mean that you're out of the ball game. Some of the finest photographs have been made by just looking at nature and trying to capture what already exists. Edward Weston (he was one of those great photographers) said, "Composition is only the strongest way of seeing." He never did fish — just nudes, bell peppers and portraits. But, these like fish and the rest of nature, are not copyrighted. Don't be afraid to reach out and explore what's around you.

Models from other materials

Other materials, like plastics and styrofoam, can also be used to form models which, like the natural organic materials, can be burned out afterwards. Plastic objects of all kinds can be used and these days there are a lot of interesting plastic shapes out there.

How about a glass Barbie® doll. Of course, you should probably try to get permission from the manufacturer before you sell something like this but anything is fair game for personal consumption. You can also use plastic models, like most of us put together as kids, as the basis for your design. Avoid

Orientate thickest sections down on your pallet.

Have appendages point in the final mold down direction.

Up in final mold

Up on pallet

Position models with flat surfaces at an angle to pallet so they don't trap air in casting mold.

Figure 62. How to orientate models for casting.

© 2000 James Kervin and Dan Fenton

ones with tiny details though, as these can be hard to reproduce in glass.

Be aware that some plastics may not burn out completely. They could leave a gummy residue. So you may want to consider test burning a small piece of the object in your kiln ahead of time. To prevent messing up your kiln, put the plastic in a small tin can (not aluminum) and place the can on the center of your kiln shelf, well away from the elements. Cover the can with a small piece of non-aluminum scrap metal and heat it up to about 1200°F to see how cleanly it burns out. If it does, go with it. If not, try making a master mold of the object and casting the shape in wax as will be described later.

When burning plastic out of a mold, be sure to ventilate the kiln and the studio well. The fumes produced when burning many plastics are very toxic and can be a serious health risk.

Preparing models for casting

The final steps in preparing your model for casting is to decide on an orientation for it in the mold and to develop the gating system for delivery of the glass into the mold as well as that to allow air to escape.

Model orientation

When placing a model in the mold, you would like all parts of the model as much as possible to point down into the final mold as illustrated in Figure 62. This allows gravity to assist in filling of the mold. To have model point down in the mold, means that they will be pointing up on your pallet since the top of the mold is cast on the surface of the pallet.

Try to position flat surfaces of the model in an off-horizontal orientation so that they will be less likely to trap air. Also, try to orientate the thickest sections of your model up in the finished mold (or down on the pallet). This prevents having to feed the glass to the thicker casting sections through thinner sections where flow will be slower.

Gating your model

Now that you know how to orientate the model, you have to work out a gating system to carry the glass from the sprue cup or reservoir, that funnel at the top of the finished mold, down into the model cavity and to allow trapped air to escape.

Since glass frit shrinks in volume during the casting process, you need to provide a fair sized reservoir to feed your mold. This reservoir should hold the extra frit needed to fill your mold as the bulk volume of the glass frit shrinks in the casting process. The top of your funnel should be sized to provide enough extra frit to allow for the approximately 30 to 50% reduction in bulk glass volume during the kiln casting process.

If the mold is an open bas relief one, this is not a problem. You just pour the glass frit into the open face of the mold and mound it up to allow for shrinkage. If this is not the case, you will need to include a reservoir and a gating system in your mold plan. This system can be as simple as a single **sprue** (channel through which the glass flows) from the reservoir to the model.

Alternately you could have multiple sprues feeding different sections of the mold from a central runner and multiple riser vents to allow air escape as shown in Figure 63.

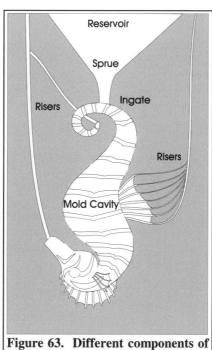

Figure 63. Different components of a gating system.

As you design this system, there are a number of principles to keep in mind. First as just stated, the cup or reservoir should be one-quarter to one-third of the volume of the whole piece. Reservoirs of wax can be cast in small paper cups, which if too big can easily be carved down. Often the reservoir will be incorporated into the design as a base for the piece as seen in Figure 64 that will only need smoothing after casting. Otherwise it will have to be cut off at the sprue and polished up. We will discuss this more in our chapter on finishing.

Sprues should be made as smooth as possible to prevent glass from hanging up in them. For the same reason, make them as short and as large as possible. Have their smallest section be at least one quarter of an inch in diameter. If they are to be removed, try to place them in areas without too much detail that you will be able to clean up later by polishing.

Attach the gating, which is usually made from wax wires and rods, securely to the model using hot wax that is dripped onto the connection and shaped with hot tools. Clay can also be used to make gating, but it is harder to clean out afterwards. If you do use clay, make sure that you clean it all out.

Smooth the ingates, those connections between the sprues and the model, and taper them slightly toward the model to allow easier removal later. Vents or risers should be attached to any features that are higher up in the model than the area that feeds it. This will allow the glass to push out the air in these areas as the model fills. They only need to be between $1/16$ and $1/8$" in diameter to do the trick.

The last step in preparing a model for casting is to firmly attach your model down to the pallet to prevent it from floating up into the plaster. If the model is of wax, this can be accomplished in a number of ways. The first is to just go around the

Figure 64. Pepper mounted on clay reservoir base.

edge of your reservoirs with an old soldering iron. You could also just get the bottom hot and stick it down or you could stick it down with low melting temperature sticky wax. If you are using a cardboard pallet, like Dan prefers, you can attach the model down with hot glue.

Replica Casting

As you work with modeling, you may develop an original model that you find particularly pleasing and which you would like to duplicate. Master molds are used for this purpose, they form a spatial negative of the original model in which multiple wax replicas can be cast. Otherwise you can imagine toiling hours, days, or even weeks to produce that perfect masterpiece only to see it melt down or go up in smoke as your investment mold cracks. While the Art police may be happily away on a fishing trip, that lawyer by the name of Murphy is always lurking in the shadows. A master mold gives you multiple chances to get your piece right.

All you need to do in order to cast wax replicas of a model is to make a master mold of the original. A master mold needs to allow for easy removal of the original model and casting of new wax models. Before starting construction of a master mold, examine your model and think where you would put parting planes. These planes will be where you will cut the master mold apart after casting it around your model. They should minimize undercuts and allow easy removal of the replica from the mold.

Try to make sure that the parting lines do not run through some point where you have a lot of detail. This is because there will usually be a little flashing at any parting point that you will have to remove from the wax casting. The best places for parting of the mold are along edges or on broad smooth areas that can be smoothed afterwards. Some models because of their geometry have very easily defined parting planes, other do not.

The kinds of materials that work best for making master molds are flexible ones. The materials that we suggest for doing this are plaster, clay, alginate, room temperature vulcanizing (RTV) rubber, Gelflex, latex, and moulage. Why do we use these materials? They are simple to work with and all will reproduce detailed images, even those with mild undercuts. Once formed, most are easily flexed and distorted to remove the model, although some are more durable than others. Let's look at each of these materials in turn.

Plaster master molds.

Plaster is a little different from the rest of the master mold materials because it is rigid and not flexible. We have already talked some about how we use plaster in modeling. It is more often used for making master molds. The simplest plaster master molds are single-piece, open-faced ones. To make such a mold, the original model must have a flat back, no undercuts, and good draft.

Draft as you remember means that the sides of the original have to be tapered to allow easy removal of the model once the plaster has been cast. Because plaster expands when it sets, the mold will compress down upon the model possibly making its removal later more difficult.

You cast a single-piece master mold just like you cast a regular waste investment mold — by gluing the model down to a pallet, building up a mold frame, coating everything with mold release, and then casting away. You do not need to add any refractory materials to the plaster or chamfer any of the edges after casting a master mold because you are not going to heat it up. You may still want to screed it though so that the wax castings you make in it do not vary in thickness from the mold not being level.

Since you will want to use the master mold a number of times, you need to construct it so that the replica can be easily removed without destroying either the model or the master mold. For this you want to make the mold in sections that can easily be disassembled. We discus this in great detail later in our discussions of multi-part molds so we will not go into it here.

Once finished with the master mold, saturate it with water and coat it completely with a release agent, such as Pam, to allow easy removal of the wax multiple. If you don't, the wax may penetrate into the plaster and attach itself firmly to the surface of the plaster so that you may not be able to remove it from the mold. You may even want to seal it with thinned down shellac or an acrylic lacquer to make it impermeable to the hot wax. Some of us had to learn this lesson the hard way. (Right Jim.)

Clay master molds

Perhaps the easiest-to-use flexible master mold making material is clay. It is good for some models, especially flat, hard medallion-like ones. You just make an impression of the model in the clay. This can be done either by making a small clay pancake and pressing the model into it or by building up the clay around the model bit by bit. The model is then removed by flexing the clay as necessary and lifting it out. A little mold release or powder such as talc if applied ahead of time will assist in removal of the model.

Master molds of clay are of rather limited use because they are easily damaged and stretched out of shape. The stretching could be used to good advantage though if you want to distort the shape of your original model. Also if your original model is rigid, you can easily make many clay master molds.

Alginate master molds

The use of alginate in making master molds is usually restricted to limited-use, short-term molds. Those of you who did not take very good care of your teeth in your youth may already have some familiarity with this material. It is the material that dentists use to make impressions of your mouth and teeth when they are making crowns, plates, or dentures. It is a natural substance that is extracted from the gelatin-like algae or kelp that we find at the seashore. You can purchase the material from dental, jewelry, or art supply stores.

As purchased, the material is a light tannish powder that, if purchased from a dental supply store, may even be flavored. (Anyone miss lunch?) The powder is mixed with water to make a mush with the consistency of creamy mashed potatoes. It must then be quickly applied to the model because it has a very short set time—on the order of two minutes. Apply it by scooping it out of the mixing container and either pouring it into a casting frame, or pressing it around the model. After application, allow 5 to 10 minutes for complete curing of the mold.

Alginate can be used to make very exact and detailed molds of you original models but because of its relative fragility, it is not the best mold material of choice for convoluted models. A lot of flexing in trying to free the model or wax replica can result in ripping or tearing of the mold. For that reason, you may only get between one half to two dozen wax replicas (depending upon the complexity of the model) from such a mold before they start to break apart. Addition of a little bit of separator compound will help maximize the number of replicas achieved.

Another limitation with alginate master molds is that they rapidly dry out and shrink. This will usually happen over about a twelve-hour time span during which they also get more rigid and brittle. Their useful life can be extended somewhat by trying to minimize water loss. This is done either by sealing the mold in a plastic bag or by storing it in water. Best results are achieved if the wax replicas are made within the first couple of hours after making an alginate mold.

Since alginate molds are so soft and flexible, you may want to cast a plaster mold around the alginate mold to give it strength. This is especially important if the model is very big.

RTV master molds

Room temperature vulcanizing (RTV) rubbers are usually two part silicone rubbers that you mix up and cast around your model. It is called room temperature vulcanizing because it does not need to be heated to cure.

An RTV rubber that works well in making master molds is Poly 74-30 made by Polytek Development Corporation. This material comes as a two part mix that when mixed together hardens to a make a lasting mold that is both strong and flexible. Unopened, the two parts are usable for at least a year from shipment. After opening, both components tend to absorb moisture from the air and should be used as soon as possible. The two parts are mixed by weight in inert containers. They are then poured into a mold frame constructed around your model.

The model could also incorporate a sprue as part of the model to assist in filling the master mold and to allow for shrinkage of the wax during cooling.

The size of the sprue depends on whether it will also be used in the refractory mold filling with frit or is only to be used only for filling the master mold with wax and is then to be removed or added to. Because of the expense of RTV, the bulk of the reservoir is usually added later. Once a master mold is cast, it is usable within 48 hours but the material really takes 7 days to reach optimum cure. These molds should be cured in a warm location (60-140°F).

After it is cured, you will need to make a couple of cuts into the master mold to get your original model out. The object is to cut into the master mold in such a way that the mold is never completely separated or that it is easy to put back together. Keeping the mold in one piece allows easier assembly when you cast your wax replicas.

To cut the mold apart, hook the rubber down to your bench with something like a bent nail. Then with a sharp hobby knife start by cutting some male-female alignment features or natches into the corners of the mold. Then slowly cut the model out of the mold trying to use your memory of where the model is to guide you in your cuts. You are trying to get as near as possible to the previously considered parting plane.

As you start reaching the model, bend the mold open as much as possible to help guide you in your last few cuts. You may also find it helpful at this point to make some cuts into the inside surfaces of the mold to make it easier to open and remove the model. This is especially helpful around delicate areas of the mold.

It is best if you can free the model without completely cutting the mold apart. The difficulty of this process depends a lot on the form of the object. A simple object like a marble is easy to do. Something like a hollow vessel is much harder and of course there are also some things for which it just may not be possible to keep the mold in one piece.

If you have any chambers in the mold cavity above the entrance sprue, you may have to drill a small hole through the RTV to the top of the cavity to act as gating for venting during wax injection. You should then insert some small brass or copper tubing like that used for hinges in making stained glass boxes to ensure that the hole stays open. Any wax can be cleaned out of the tubing between castings by running a hot needle down it. The RTV master mold is now ready for use.

Gelflex

This is another reusable master mold material. It is a polyvinyl material, which can be used on a wide variety of modeling materials many times without needing a release agent. Depending upon the replica material being used, the mold should be good for multiple casting cycles. Unlike many master mold materials, it can also be retouched. It is available in two different grades hard and soft from Schell studios.

Before using it, the first thing you need to decide is if you need sealants or release agents on the model. Plaster models should have any paint removed and be sealed with shellac. Stone models should be sealed or at least soaked in water overnight. Clay, metal, glass, or china models need no sealants. Wood models should be sealed but even then will often outgas during casting producing mold flaws. The only model materials that really require release agents are metals. A silicone lubricant oil is the ideal release agent for them.

Next set up the model in a mold frame like those used for investing. Secure it and the walls to a glass pallet. Use hot glue or clay to secure them. The walls of the mold frame should be made of impermeable materials like glass, formica, plexiglas, linoleum, or sealed wood.

You must next decide which grade of Gelflex to use: the harder blue-colored one, the softer natural-colored one, or a mixture of the two. Natural will pick up greater detail but blue is much more durable. Durability is more important the more replicas you plan to cast and the larger your model.

Now cast your master mold. Melt the Gelflex in a old electric deep fat fryer not a crock pot. This should be one that you will never use for cooking purposes. It should have the capability of heating to 325°F and a working thermostat to prevent it from getting much hotter than this. Going higher will shorten the life of the material drastically. Use a thermometer like that used for candy to accurately monitor the temperature of the melt.

Melt only as much material as you need for the project or about 10% more. It is advisable to heat it in a ventilated hood or space. If you do not have enough ventilation consider wearing a respirator rated for chemical vapors.

Set the thermostat of the fryer for 325°F and heat up the Gelflex. Check the thermometer occasionally to keep track of what is happening. As it starts to melt, stir it gently with a heavy glass stirrer. This will help eliminate air from the melt. As it melts, some vapors will be released. Covering the melt can minimize vaporization.

After the Gelflex is completely melted, turn off the fryer and let it start to cool. If the fryer has a removable liner then remove it from the pot with gloved hands and set it aside. When the Gelflex temperature has dropped to anywhere between 266 and 302°F, it is ready to pour. Using the lower temperature reduces air in the mold and speed of set.

Like investment, pour the Gelflex into a corner of the mold frame and not directly over the model. This will reduce the chances of trapping air against the model. After the master mold has cooled, cut it open and use it like any other master mold.

Once the mold has used up its useful lifetime, the material can be recycled. Start by cutting up the mold and washing the pieces off with hot soapy water. Do not use any organic solvents on them as this will denature the material. Clean out any foreign material and then just remelt it.

Latex rubber master molds

Liquid latex rubber compounds are another good master mold material. Don't confuse latex mold material with latex paint. You won't find this at your hardware store. The latex for mold making can be found at sculpture supply, plastic's supply, and hobby craft stores.

These solutions are available from a number of different manufacturers. We have found that you don't have to buy the most expensive kind, because the cheap kinds seem to work just fine. "Mold-It" from a hobby craft store has worked the best. Another good one is D & S 74 from Douglass and Sturgess in San Francisco.

Latex is the natural rubber sap from a rubber tree. It has been suspended in a base made from a mixture of water and ammonia. Its normal shelf life is about nine months if kept tightly sealed in an airtight container. A metal foil sealed lid is desired if available, because it will help prevent the latex from drying out. If it does dry out, you should dispose of it and get some more.

Store your latex solution in a cool place. But not so cold as to let it freeze because that will ruin it. If exposed to light, the rubber solution will take on a pink tint over time. Don't let the color change bother you because that will not affect its properties. If the change in color does bother you, then you need to store it in a dark spot.

Like the RTV rubbers, latex's can reproduce great detail from a model. On the down side, they are a less rugged than the RTVs but on the other hand that means they are much more flexible than the RTVs. Latex master molds are so flexible that they can even be turned inside out. They are also cheaper which always helps. One thing that they do not accommodate well is inside passages that go through the mold.

Latex will react with petroleum or sulfur-based materials. For this reason, it can not be used over plastecine clays. If you want to make a latex master mold from a plastecine clay model, you will have to seal it with thinned down shellac.

Construction of a latex master mold is simple. Latex is applied by painting it onto the original with a cheap disposable brush like the ones used for applying soldering flux. It's sometimes easier to throw the brushes away than to try to clean them up afterwards, especially if you forget to pretreat them. In a pinch, the brushes can be cleaned with mineral spirits. Latex is used as follows:

1. Place your original model on some base surface like glass or linoleum, which will provide for easy release afterward.

2. Coat the base and the model with a light coating of some mold release compound. Murphy's oil soap or Pam spray both work well for this. Release agents are absolutely necessary for any application with a model that has any appreciable porosity.

3. Ready a small soft bristled brush for application by precoating it with liquid soap. This helps prevent the latex from sticking to the brush. Remove any excess water

Replica Casting

Figure 65. First layer of latex painted on models.

before dipping the brush into the latex solution.

4. The latex can generally be used straight from the container. If you feel that it needs stirring, do so gently and slowly so as not to entrap air. Do not for any reason shake the latex solution. This is sure to mix air into it.

5. Apply a light coat of latex to the base surface in an area at least 1" around the model to anchor the mold to the base.

6. Apply a light coat of latex to the entire model taking care to cover all detail. The detail that you will achieve in the master mold depends on the smoothness of this first coat. You may even want to thin out the latex a little for this first coat with distilled water. If you do trap any air bubbles in the first coat, blow on it to release them or prick them with a pin.

7. Drying of each coat is usually fairly rapid and can be accelerated by blowing on it with a fan or by heating with sunlight, a hairdryer, an oven, a kiln, or some other heat source. Use a low heat setting. The popular question is how do you know when a coat is dry? Usually when it dries, the latex changes tone and becomes slightly less translucent. Of course this will vary depending on the exact product used. If you wait 60 to 90 minutes, you can usually be pretty sure that it will be dry.

8. After each coat, soak the brush in soapy water to keep it clean and precoat it again with detergent prior to adding each successive coat. Comb out any latex left on the brush when you are done for the day.

9. As soon as the first coat is completely dry, at least to tack free but never over 24 hours, add a second coat of full strength latex solution all over the model and some extra to the apron on the base extending the apron slightly on each coat.

10. Brush out any pools and drips of latex so that each coat will dry completely. If allowed to pool, it will form a skin and the interior of the pool will dry very slowly. Be sure to replace the lid on the latex container between applications. Prolonged air exposure can greatly alter the material's performance.

11. For a single use mold, you could probable stop after about four coats. For a semi-permanent mold to make many replicas, you want to apply up to 8 or 16 coats. After adding your last coat, allow the latex 24 to 48 hours to cure completely. If you want to really speed up the drying process you can always use a hair drier.

12. Strength can be added to a mold by embedding a cloth mesh backing or cotton flocking into the mold after about 6 coats and continue on with about 8 to 10 more coats. The flocking is mixed with latex in a ratio of two parts of flocking to three parts of latex adding the filler to the latex to make a paste. Do not make the paste so thick that it will not go on smoothly or dry properly and only make up as much as you can use right away.

13. For a mold that will see a lot of use or for larger molds, you may also want to cast a two-part plaster mold around the latex mold to give it strength during filling.

14. After the final latex coat is completely dry, slowly and carefully remove the latex coating from around the model as was explained for the RTV mold.

If you find a snag, work it out carefully. Snags, areas sticking to the model, occur because of a gap in the mold release or from a thin spot in the latex coating. If you end up with a hole in the latex, it is usually possible to patch it by painting on some more latex. Snags may also be an indication that the shape that you are trying to duplicate is much too complex.

Like with RTV molds, it may be necessary to add vents to your latex molds to ensure complete filling when injecting them with wax. Prior to injecting wax into a latex mold you may want to close cuts made in it by painting on a little latex along the cut like adding glue and letting it dry. This will make the latex master mold more stable during wax injection.

To store the latex rubber mold for long periods of time, you need to keep it in a cool place. Do not wad it up. If you cast a backing mold of plaster, store it in that. A mold that is well cared for can last up to 20 years.

Moulage master molds

Moulage is another reusable material like Gelflex that can be melted and used to form a master mold. Its main benefit is its low melting point. This allows it to be used for making molds of body parts without burning your model. Its being nontoxic also helps.

The material when purchased from your vendor has a moist rubbery texture and resembles large curd cottage cheese. It should be stored with a little water in an airtight container like a plastic bag. This should be kept in a cool place. If treated right, the material can be recycled and used over and over.

Figure 66. Melting moulage in double boiler.

Moulage should only be used over non-porous models. Any porous models will have to be sealed with something like thinned down shellac. Since moulage molds are not very durable, you may want to use it to cast a plaster positive and a subsequent rubber negative mold if you are going to want to make a number of wax replicas of the subject.

To prepare your moulage for use, determine how much you need. One pound of moulage is about fifteen cubic inches of material. Melt it in a double boiler. This double boiler can be made from porcelain, glass, or stainless steel but not from aluminum. <u>Moulage should never be used with aluminum</u> because they are not compatible. Heat the moulage until it becomes a thick and creamy liquid.

To use it on live models, let it cool to about 110°F. To cool it quickly, replace the hot water in the bottom of the double boiler with cold. Stir it to maintain an even temperature and prevent solidification on the bottom of the pot. After it gets to about the right consistency, replace the cold water with warm water for longer pot life of the material and it is ready to apply.

After a mold has exceeded its useful life, it is time to recycle the material. Clean off any foreign material, wash the mold off with water, and cut it up into small pieces. Store those pieces like the new material, wet in a tightly sealed plastic bag. The material can be used well over 100 times without any noticeable loss of capability.

Life modeling

Life modeling refers to making molds from live subjects. As just mentioned, the ideal material for this is moulage because of its low melting temperature and its excellent ability to pick up detail. Alginate is also a good choice but you have to work quick. Many people also use plaster bandage material if a less detailed mold is acceptable.

For this discussion, we will discuss how to cast a mold of a face first with moulage and then with plaster bandages. The reason that we discuss doing a face is because there are many more features to deal with and if you can handle it then you can do easier body parts like hands, feet, and other more erotic features.

Before you start, sit down with the subject and explain the process to them. Tell them that the process will take up to 45 minutes and that they will have to remain still during the whole process. Explain that their whole face will be covered and that they will be breathing through two straws inserted into their nose. They should not be afraid.

The first thing to do is make your model comfortable. Position him or her such that the view that you want is more or less positioned up or at most at a 20° angle. The most comfortable position is probably laying down with their head on a pillow. To avoid the moulage from running all over the

Replica Casting 69

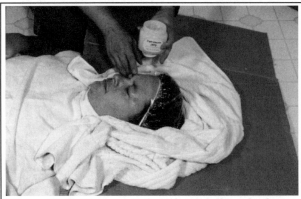

Figure 67. Getting ready for life modeling of a face.

place, you may want to cut a piece of cardboard to fit around the face as shown in Figure 67. It should fit fairly close. Towels should also be wrapped around the subject over or underneath the cardboard to help contain the moulage.

Next you want to prevent the moulage from getting into the subject's hair or ears. Ears should be plugged up at least with cotton. If they are exposed, you might want to use a better seal like earplugs. Realize that if the mold wraps around the ear, release may be complicated and painful.

Any exposed hair should be slicked down with a heavy coat of Vaseline so that it does not get embedded in the moulage. How much Vaseline you will need depends upon the subject's hair length. You may want to actually avoid this issue by having the subject don a shower cap or a swim cap. You can always rough in hair on the replica using modeling tools.

Now melt the moulage in your double boiler. Cool it back down to about 110°F or until the subject can tolerate it. Test a small amount on your wrist first like you would a baby bottle. Then check it out on a

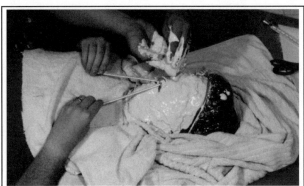

Figure 68. Applying moulage to a model's face.

sensitive spot like under the chin on your subject to see if it causes any discomfort. If it is cool enough, you are ready to make the mold.

Apply the moulage with whatever is comfortable— fingers, spoon, brush, etc. You have to apply the entire mold at one time because a second layer will not adhere well if applied over a cooled layer. Build up about a half-inch layer over the whole face. Start at the top of the face and work your way down to the chin.

Leave the nostrils until last. Before covering the nostrils, insert straws into the nostrils so the subject can breath after the moulage is applied. You may want to cut the straws about three inches long so that they are easy to handle. Let the subject have control of the straws so that they will not get freaked out. They will also be calmer if you talk to them or play their favorite music. Smooth out the surface of the moulage as shown in Figure 68. It will take about twenty minutes to set up.

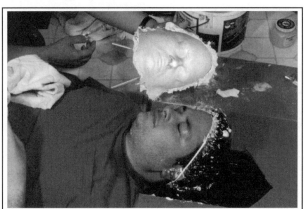

Figure 69. Finished moulage mask being removed.

Since the moulage will not have much strength, you will want to cast a plaster shell over it. So while the moulage is starting gel, mix up a batch of plaster or get some casting bandages ready to lay up over the mold. Once the moulage has gelled, apply the plaster at least one half of an inch thick. Apply the plaster after it has reached the consistency of pudding. Bandages only have to be three or four layers thick. Smooth either out and let them set.

While the plaster is setting, it will get warm. Make sure that the subject understands this so that it does not frighten them. This is also the reason that plaster is not the right material to use for life modeling. Let the plaster set up. If you want, you can use a hair dryer on cool to speed up the process.

Once the rigid outer mold has set up, it is time to remove it from the subject. Either lift it off or have the subject bend forward and let gravity pull it off. Dry off the interior with a soft towel and you now have a mold of a living person ready for replica casting.

Figure 70. Applying plaster bandages to a model.

Making a living mold from plaster bandages is much quicker and easier to construct but captures much less detail. Prepare the subject by slicking down their hair and wrapping them in a towel as for making a moulage mold. You do not have to worry about as much mess. You do not need the cardboard to keep the bandages from getting all over the place.

Next cut up the plaster bandage into about six to eight-inch lengths. This makes them easier to handle and it is easier to do now then after you start getting things wet. Now get the subject ready. Warn then that the bandage will warm up, but since you are not going to apply too much plaster that it will not get too warm. Put them in a comfortable position

After you have the subject prepared. Dip some of the plaster bandages in water. Then lay them on the subject getting it to conform to the face. Then grab the next piece and place it. Move around and cover the whole face as in Figure 70. Since the plaster bandage is so strong, you will only need to make it three or four layers deep. Work around the nostrils like before but you do not really need the straws this time because the plaster bandages will not flow.

After you have completed laying up the mold, you will only have to wait about three or four minutes for the plaster in the bandages to set up. After it has set up, it can be removed from the subject and trimmed with a scissors. The holes for the nostrils can be covered with some plaster bandage and the mold thickness can be increased if desired.

Using master molds to cast replicas

Now that you have your master mold, you can make as many wax replicas as you want. Generally speaking, the way that you do this is to inject molten wax into the master mold. Jewelers use hydraulically powered wax injectors for this operation. A simple kitchen turkey baster or the wax extruder that we discussed for wire manufacturer are both good alternatives, especially if your model is not very big.

Wax will build up on the inside of a baster over time and can plug it up. You will have to heat it up with the wax to try and break any logjams. Always preheat your injector before use and empty it after use.

Special casting waxes have been developed for injecting wax replicas that melt and set quickly as well as being more fluid. These waxes when set are fairly hard and slightly brittle. You can use these waxes or some other hard one of your choice.

If you do not have an injector, you will have to either pour or paint the wax into the master mold. Pouring wax to fill a mold is in many ways very similar to injecting it. You just pour it into the mold and let it cool. Since you are not doing it under pressure, your wax should be a little hotter than what it would be if you were to inject it to ensure that it fills the mold.

Heat the wax to temperatures between 100 to 250°F before injecting it. Then let the hot wax to cool back down to the solidus point, the temperature where a skin just starts to form on it and inject it into the master mold. If too hot, the wax may dissolve some of the mold release and stick to the mold. This can

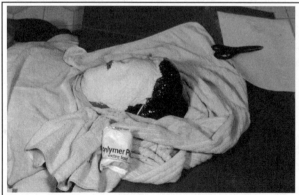

Figure 71. Completed plaster bandage mask.

Replica Casting

ruin some master molds and will surely scar the wax casting during mold removal. Another problem with injecting your wax too hot is that it will shrink more as the bulk cools down and this can cause cracking or other imperfections in the replicas.

The cooler the wax is when you introduce it into the mold, the less it will shrink or stick to the mold. Although it should be obvious that if you let it get too cool it will not flow into the mold very well.

One way you can tell if the wax is cool enough to pour is watch when a slight film starts to form on the wax. Then wait about 10 minutes and sticks your finger in it. If you end up performing the "hot wax dance," you can be pretty sure that it is still too hot.

We do not really recommend this practice unless you have chrome-molly fingertips. You may want to try a candy thermometer instead. Be very careful when injecting the wax that you do not spray it around and burn yourself. This can be a problem if you push the injector against the back of the mold and back pressure builds up as the master mold fills up. Take it easy.

Once you have injected wax into your mold, give it a little time to start to set but try to remove it from the mold while the wax is still flexible to make removal as easy as possible. Some people find it easier to cool the wax until it is hard before removal. This can be hurried by putting the mold into the freezer for a half-hour or so.

Casting wax into plaster master molds is not a delicate process as it is with latex master molds. It is still a good idea to let the wax cool to the solidus point before pouring, but that depends upon the shape. A more complex pattern with small passages might be more easily poured if the wax is a little warmer but you stand the danger of it not separating well. The plaster will absorb heat from the wax while the latex acted as more of an insulator. If you happen to forget the mold release, don't panic, just steam the wax back out and try again.

Clay master molds are usually waste molds so do not worry about trying to save them. Level them and block them up with some more clay as necessary and you are ready to go. Because they are waste molds, they do not require any mold release. Just inject your wax and let it cool. Then peal away the clay from the replica and you are done. If you want another replica, then repress the clay onto the original model and you are ready to go.

For alginate and moulage master molds, the walls are usually thick enough to be self-supporting and if they are not you will have probably cast a plaster mold around it to support it. If need be, you can set them up in sand like the more flexible latex molds that will be described shortly. No release agent is absolutely required but it will prevent sticking and will allow casting more wax replicas from a mold. After the wax has hardened, flex the mold as necessary to release the replica.

For a thick RTV rubber box mold, clamp the mold together between two flat plates using spring or C clamps. Again, no release agent is absolutely necessary in the mold for casting your wax models but a little silicone spray or cornstarch would not hurt. Then inject the wax, allow it to cool and remove it by opening the mold.

Figure 72. Model, latex mold, and wax replica.

Casting wax into a latex mold needs to be done a little more carefully but it is not too different. The latex is flexible and therefore needs to be supported by a cradle mold during injection. Otherwise it may slump and distort or worse yet, roll over and dump hot wax all over. A box of sand is usually just the ticket to serve as a cradle. Use 80 or 100 mesh silica sand. Avoid "sand-box" sand because it is too dusty and presents a silicosis danger. (Talk about dangers on the playground.)

Press the latex master mold into the sand, level it, and adjust to shape. This whole process may be made easier if you slightly dampen the sand. It is possible to creatively alter the shape at this point but most artists choose to be true to the original model. When the latex is set up in the bed of sand, spray the inside of the mold with a little mold release and inject the wax. Once cool, roll the latex mold off the wax replica.

When trying to separate a latex master mold from an original wax model, it often helps to try and harden the wax by sticking it into the freezer for an hour or two. Then when ready to cut it free, briefly warm the latex with a hair drier, but not too much. You are just trying to take the brittleness out of the latex without softening the wax.

One thing that you may notice after removing the wax replica from a master mold is that it will be slightly smaller than the original model. This is because the mold material shrank around the original model and the melted wax shrank while solidifying in the master mold. This will probably result in a copy that is about 5% smaller in volume than your original. This new wax model is now ready to invest for frit casting. Figure 72 shows such a series of components. See if you can notice the difference in size of the original model on the left and the final wax replica on the right.

Casting a hollow replica

If your model is fairly large, you may not want to have a solid wax replica. In this case you may want to make a hollow wax multiple with a method similar to slip ceramic casting. To do this, first paint a layer of hot fluid wax into your master mold, paying special attention to any areas of great detail. It would be ideal if this was a little higher melting temperature wax than that which you use next. Areas with sharp angles may need to be built up even thicker as the next step may erode some of this away. Now close up the master mold and fasten it shut. Use a plaster backing mold if you have one.

Next pour some hot wax into the mold that has been cooled back down to just barely over the solidus temperature so that it will harden quickly. Roll the wax around in the mold for a few seconds if it is not enough to fill the cavity. Then pour out the excess wax.

The Italian masters had a trick for judging when to pour out the excess wax from a mold and end up with the thickness that they wanted for the wall. They would heat up about twice the amount of wax needed to completely fill the mold. They would then pour half of it into the mold to fill it and would set the rest aside to cool. After the wax in pan had cooled enough to form a skin over its surface, they would keep a close watch on the thickness of the solid wax that was forming on the side of the pan. When it was about the thickness of what they wanted inside the mold, they would pour the excess wax out of the mold. What was left was then about what they wanted.

After the wax has completely hardened, you can pull back some of the master mold to determine if the wax is really thick enough. You would like it to be at least ¼" thick to hold up under the weight of the mold material during investing. If not, repeat the process until you reach the desired thickness.

If you are using a number of copies of an original model in one casting such as a school of fish, one thing that you can do is to flex and shape each of them into a slightly different pose. They can then be grouped or cast singly.

Finishing a replica

You may find that you also have to clean up your wax replica in the parting region of the master mold where the pieces come together. It may have flashing or slight imperfections in this region. These can usually be removed by scraping across it as discussed earlier with a sharp X-Acto knife. You may find it beneficial to cool the wax replica before doing this.

You can also touch them up using some of the other techniques that we discuss for wax modeling in the last chapter.

Investment Materials and Formulations

Success in pâte de verre and kiln casting depends a lot upon how strong your casting molds are which in turn depends upon which investment materials, and formulations you use. The term investment is derived from a term once used to describe special apparel such as cloaks. Thus, when a model was "cloaked" with a mold, it was said to have been invested and the material came to be called investment.

Whether you should formulate your own investment or purchase a commercially prepared one usually becomes a matter of choice based upon convenience, availability, and cost. In most cases, a commercially prepared investment will have more consistent results because the manufacturer has facilities to carefully control its manufacture. Also since they are a large company, they can purchase the components in enough bulk that they may be able to sell it to you at a price below that with which you could formulate it yourself.

On the other hand, you may be located a great distance from a manufacturer or distributor of investment which would require considerable shipping cost to purchase it. In this case, you may be forced to formulate and mix up your own investment. So let's learn more about the materials used in investment formulations and give you the option to pursue either option.

Investment materials

Before giving you some basic investment formulas to work with, it is best if you understand what the functions of the different formulation components are. Otherwise, if you develop problems, you will not have the basic knowledge to troubleshoot what's wrong and what you have to do to fix it. With this basic knowledge you will have many options. When you get a formulation from someone that works well for them but not for you, or you have a formulation that works well with one kind of glass but not with another, you should be able to figure out how to modify the formulation to suit your purpose. Although again you may decide to avoid this issue entirely and choose instead to work with a commercially formulated investment.

Even if you decide to purchase your investment, you should understand their composition. There are three basic components to an investment mold formulation:
- binders
- high temperature refractories
- property modifiers.

Each component performs a different function but exactly which function a material is performing can sometimes be a little blurred in practice. To confound things further, some of the materials that we will discuss may also be available in multiple forms.

Binders

Binders are those materials in an investment formulation that bind the other components of the formulation together. The most common ones used in kiln casting are gypsum plasters and cements. You can think of them as the mortar that holds all the individual bricks together to make a wall. In the simple investment formulation that we presented earlier, the binder was plaster.

Each binder has its own operational temperature regime. As you exceed it, the binder loses strength. About half of a investment formulation typically consists of binder. As you decrease the binder fraction below this value, the investment becomes weaker. A list of typical binders used in glass investment formulations and the temperatures regions in which they break down is given Table 6.

Table 6. Breakdown temperature of common casting binders.

Binder	Breakdown Temp
Gypsum plaster	1300-1500°F
Hydrocal cement	1300-1500°F
Hydroperm cement	1400-1700°F
Portland cement	1600-1900°F
Colloidal silica	2300°F
Colloidal alumina	2300°F
Calcium alumina cement (fondu)	2800°F

Binders use three basic mechanisms to hold the other components in the investment mix together.

- **Chemical bonds** are the strongest mechanism that can form between binder molecules and those of the other components in the investment formulation.

- **Molecular attraction** is a slightly weaker mechanism between the binder molecules and those of the other components. An example of such attractive forces is electro-static forces.

- **Physical compaction** or trapping of the other components within a matrix of binder molecules is weakest mechanism.

Let's look at some typical binders in detail.

Gypsum plasters and cements

Gypsum (calcium sulfate) plasters and cements are basically ground up gypsum stone (alabaster or selenite), which has been quarried, crushed, screened, pulverized, and is then heated to around 350°F (**calcined**) to drive off both the physically attached water and the chemically attached water molecules of crystallization.

This changes gypsum from calcium sulfate dihydrate ($CaSO_4 \cdot 2H_2O$) to calcium sulfate hemi-hydrate ($CaSO_4 \cdot \frac{1}{2}H_2O$) and with more heat to calcium sulfate anhydrate ($CaSO_4$). The proportions of each of these in a particular will determine a particular plaster's properties. Some plasters will have more of one and some more of the other. As an example, Plaster of Paris is almost pure calcium sulfate hemi-hydrate. The ratios of each in a plaster are a result of the different temperature and pressure conditions under which the calcining is done.

There are even more differences than just the ratios of the three different forms. Calcium sulfate hemi-hydrate also comes in two different forms α and β. Which you get depends upon how you calcine the gypsum. If you calcine it under pressure in a humid environment, you will get α. This form has a nice even crystalline structure. If you calcine it at atmospheric pressure as is usually the case because it is cheaper, the water gets driven off as steam. This breaks up the crystals into fragments which is the β form.

The difference in crystal structure changes the strength, the set time, and the amount of water needed to rehydrate the crystals. The β form is weak, sets faster and requires more water. And of course as before you can get a whole series of different plasters by varying the ratio of these two forms.

After calcining, the gypsum is reduced to a soft material that is easily milled to produce a fine homogeneous product. The end product is thirsty and wants to return to its stone state. When mixed with water and allowed to set, gypsum plasters and cements will form a produce without grain, hardness variations, or lumps. They differ in how much water is required to form a workable product.

Plasters are combined with 65 to 160 pounds of water per 100 pounds of plaster to achieve a good pourable slurry while cements are mixed in the range of 22 to 45 pounds of water per 100 pounds of cement. They both are very hydroscopic, i.e. soak up water out of the air, off your hands, or whatever. Any humidity contamination will not be immediately apparent in the powder but it will influence setting properties.

Because of this, you have to keep them in a warm dry place away from water sources like damp floors. The primary gypsum plaster and cement manufacturing company, **US Gypsum (USG)**, recommends that plasters or cements should normally not be stored for longer than 90 days to retain their optimal properties. But if kept dry, they can be stored for much longer periods.

We recommend that you keep your plasters and cements sealed in plastic bags closed with strong ties at the very least, or better still get some of those plastic buckets with tight fitting lids used to store

Investment Materials And Formulations

Table 7. Properties of typical gypsum plasters and cements.

Gypsum Product	water added (% of dry mix wt.)	Set time (min)	Dry density (lb/cu ft)	% Set Expand	Compressive strength (psi)
No. 1 pottery plaster	70	27-37	69.0	.210	1,800
No. 1 molding plaster	70	27-37	69.0	.200	2,000
Plaster of Paris	70	27-37	69.0	.200	2,000
No. 1 Casting plaster	65	27-37	72.5	.220	2,400
Pottery plaster	74	27-37	66.0	.190	1,800
Hydrocal cement	45	25-35	90.0	.390	5,000
Hydroperm cement	100	12-19	<40	.140	-----
Hydro-Stone cement	32	17-20	119.4	.240	10,000
Ultracal cement (30)	38	25-35	99.0	.080	6,000

flour. These can often be found in the trash behind bakeries. You can also find them for sale at some hardware stores.

Transfer material from your larger strong container to the smaller buckets for use in your mixing area. This is done in order to try to minimize air space over the stored plaster or cement. Doing this reduces the investments exposure to moisture from the surrounding air. Be careful not to get water into your plaster or your mix may seem like it has pebbles in it. If you do accidentally get water into your small plaster container while mixing up a batch of investment, at least you have not lost a lot of it.

When mixed with water, gypsum plasters and cements slowly chemically change back into their original form, a rock. As they harden, they do not really dry, they set. Most of the water is still there but it is incorporated as part of long gypsum crystals. By changing the particle size and drying procedure, plasters and cements with different performances are produced.

The different gypsum products that you can purchase may also have materials added to control setting expansion or time. For example materials such as alkali sulfates or chlorides are added to accelerate set while ones such as citric acid, sugars, borax, gelatin, or starch are added to retard it and the addition of either will usually reduce setting expansion.

The different gypsum plasters and cements vary in strength, set time, expansion during set, shrinkage during firing, water requirements, and pour characteristics. Even so they can generally be substituted for one another. Doing this may require slightly different amounts of water be added to achieve the same degree of dry strength. This is not true of the gypsum cements as you will find out. Some of the commonly available gypsum plasters and cements are listed in Table 7 with a few of their typical physical characteristics.

The strength of a plaster or cement is the result of the development during setting of the matrix of numerous needle-like calcite crystals, which become tightly interwoven. It is the density of this matrix that dictates its strength. Dilution of the mix with more than the desired amount of water results in a final matrix of crystals that are further apart from each other than desired, thereby weakening it.

That is why the optimum amount of water to be added to the mix, usually referred to as the normal consistency (amount of water by weight to 100 parts of plaster to achieve a standard fluidity), is so important. This standard fluidity is that which will only let a fifty-gram rod with a nineteen-millimeter diameter penetrate thirty millimeters when release under a controlled spring loading.

Varying from normal consistency will affect the performance of the investment. As an example of this, see the data in Table 8 which lists how setting time, compression strength and final dry density of a typical plaster varies with the plaster to water ratio. The amount of water a plaster is mixed with also affects other properties besides strength such as: chip resistance, durability, density, and surface permeability.

Of course, there are other things that can affect the performance of your investment. If not stored properly, your plaster or formulated investment mix can rehydrate by pulling moisture out of its surroundings. If it has done so, it may not perform well anymore. This can cause it to take longer to set up because much of the recrystalization may have already occurred. This reduces the amount of interleaving of the needle-like crystals and thus the strength of the mold, resulting in increased mold cracking. Molds made from partially hydrated plaster many times will also be rougher to the touch because of clumping of the plaster.

Three ways of determining if you have a problem come to mind. The most accurate is to compare weights of equal volumes of your suspect plaster and a known good plaster or "hot" mix. If the weight of the suspect material is more than 20% greater than the hot mix, get rid of that old stuff. Alternatively you can compare the setting time of the two plasters. If the suspect material takes more than a couple extra minutes over the good stuff, it is most likely bad. If either of these sounds like too much work, then try pinching some of the suspect plaster between your fingers. If it retains its shape, it is probably usable. If not, then it it is probably not worth keeping. Trying to save money by using bad plaster is just not worth it.

Us Gypsum actually makes more than thirty different plaster and cement formulations. A description of some of the commonly available ones and their properties follows:

USG #1 pottery plaster is a good general purpose plaster. Its strength and easy flow characteristics account for its popularity. It is generally recognized as the industry standard against which other gypsum plasters are judged. It is mixed with 70 parts of water per 100 parts of plaster. Pottery casting plaster has polymer and synthetic fibers additions that improve its performance in solid sculptural or slush casting applications. These increase the chip resistance and impact strength of the mold. It is mostly used by the ceramics industry for the construction of slip casting molds.

USG #1 moulding plaster, sometimes referred to as "soft plaster" or "Plaster of Paris" (so called because the raw material was once mined from the hill of Montmartre in Paris) is the softest and most porous of the different varieties. It contains no surface hardening agents and is thus the easiest to carve, use for modeling, or make molds that you want to carve on. Its nominal consistency is also 70 parts of water per 100 parts of plaster. Anything in the range 67-80 parts of water works well. It reproduces intricate detail well.

USG #1 casting plaster is another widely used utility plaster. It mixes easily and is slightly harder and denser than molding plaster as well as having better chip resistance. It contains small amounts of hardening agents to achieve harder surfaces and other additives to give it smooth working capabilities. Its extra hardness results in decreased permeability, which is useful in normal model building because it reduces the amount of paint the plaster soaks up. For this reason, it is also a good plaster for making wax master molds. It is mixed to a nominal consistency of 65 parts of water per 100 parts of plaster.

USG Hydrocal is a white gypsum cement. It has a much greater setting expansion (almost twice) that of the gypsum plasters which is also the greatest of all the gypsum cements. This expansion occurs uniformly in all directions and is controlled through the amount of water with which it is mixed. This greater expansion facilitates removal of models after complete set which is usually anytime after about two to three hours from initial set. Hydrocal is fairly easily carved and will bond to additional pours after it is completely dry. It must be worked quickly because it has a fairly short period of plasticity and hardens rather rapidly. It is mixed with 45 parts of water to 100 parts of Hydrocal. There are actually a number of Hydrocal formulations available that differ mainly in strength and set expansion.

Hydroperm is another gypsum cement that is often used in the construction of molds for metal foundry casting. It is formulated with a foaming agent that causes it to produce a uniform matrix of bubbles throughout

Table 8. Effect of plaster-water ratio on some properties.

Plaster-water ratio (by weight)	Setting time (min)	Compression strength (psi)	Dry Density (lb/cu ft)
100/30	1¾	11,500	112.7
100/40	3¼	6,750	96.6
100/50	5¼	4,500	84.4
100/60	7¼	3,250	75.3
100/70	8¾	2,500	67.6
100/80	10½	1,800	61.8
100/90	12	1,400	56.7
100/100	13¾	1,000	54.1

the mold forming a lighter, low-density mold. The size and distribution of these bubbles can be varied by changing mixing parameters. Once set up, the bubbles result in an interconnected cellular structure in the mold that makes it very permeable to gases. This is a desirable property for foundry applications since it prevents trapping of air in the mold as it fills and allows the molds to dry much faster. Its formulation also includes some refractory materials.

Hydrostone and **Ultracal** are two other gypsum cements that we do not usually use because they set up too hard for mold making. You can end up breaking the casting as you try to remove the mold. They can be put to good use though in making cradle molds to reinforce your casting molds or as a partial addition to strengthen molds if you are plagued with cracks.

Many glass artists will refer to any of the gypsum plasters or cements as plaster without recognizing that there is really any difference between them. Sometimes we are no different, although you would think that we would know better. As a general rule, we tend to use white Hydrocal cement for a binder in our investment formulations.

Cements

Besides the gypsum cements, there are a number of other cements that you may want to use in your work as partial additions to strengthen your investment formulation.

Portland cement is most commonly used in construction. In mold formulations, it is used with plasters to modify strength or set time. Unfortunately Portland cement becomes chemically activated during casting operations and would readily stick to the glass if it were used as the primary binding agent for high temperature (over 1200°F) applications. Portland cement can be used as a backup insulating refractory. It should not be exposed to direct contact with glass or flames.

Calcium alumina cement, commonly known as Fondu (a brand name), is a high temperature (up to 2500°F) binding agent commonly used in castable refractory formulations that need to withstand direct exposure to flames and furnace gases. Up to 15% by weight can be added to an investment mix to increase a mold's strength at high temperature. One down side to using it, is that its strength makes the mold harder to remove from the glass after firing. When added in amounts greater than 4%, small amounts of potassium sulfate are required to facilitate quicker setting of the investment. Calcium alumina cement is the binder used in the lightweight castable refractories manufactured by A. P. Green. Glass fusers also sometimes use it in the construction of slumping molds.

Colloidal alumina is a liquid suspension of fine alumina particles. It is used as a binding agent for tightly packed olivine sand when doing sand casting. We glass fusers sometimes use it for rigidizing fiber paper and blanket in slumping mold manufacture. Although it can air set, colloidal alumina should really be fired to about 1100°F to reach its optimum strength. It does not become activated at high temperatures so it will not stick to the molten glass. This means that it can also be used as a good glass casting mold release. Its addition to an investment formulation will strengthen it and reduce glass sticking.

Colloidal silica has similar binding properties and uses as colloidal alumina. It is also used by many glass fusing artists to rigidized ceramic fiber molds. Unlike colloidal alumina, colloidal silica can become activated at high temperature causing it to interact with or stick to molten glass and therefore does not work as a mold release. It needs to be fired to realize its optimal strength but it only needs to be fired to about 800°F. If you use it as a binder you may find that a kiln wash mold release is required to prevent glass from sticking to the mold.

Refractories

The next major investment formulation component, refractories, are materials that are added to the formulations because of their high temperature stability. As materials go, they tend to be stable both dimensionally and chemically at high temperature. It is for this reason that refractories also form a major portion of an investment formulation. Typical refractory materials added to investment mixes are: silica, diatomite, alumina hydrate, zirconia and olivine sand. These microscopic refractory particles influence the shrinkage of the binder matrix during firing. They do this by acting as a stable backbone to which the binder matrix can cling as it shrinks.

In adding refractories to a plaster-based mold formulation, it is important to add a material or materials that provide a range of particle sizes. This allows you to more closely pack the refractory particles in the investment mix so that they form a stronger and denser matrix. Then when the plaster starts to shrink at about 1200°F, it can grab tightly onto this stable matrix resulting in increased mold strength and decreased shrinkage. This works well until about 1400-1550°F when the plaster starts to decompose and lose its strength.

Silica-based refractories

Silica may be added to an investment formulation as either a silica sand (120 mesh and coarser) or a silica flour (220 mesh and finer ground from flint). Particles larger than 80 mesh can be problematic because of their tendency to settle. This problem is worsened if vibration is used to try and release trapped air in the freshly poured investment.

Silica can cause problems in high temperature molds for two reasons. First, because of its relatively low melting temperature, it can become activated and stick to the glass. Second, being a finely ground form of flint, it goes through a quartz inversion at approximately 1060°F. At this temperature, the crystals rearrange themselves with about a 2% volume change causing the silica to expand during heating and contract during crash cooling of your kiln. This can cause cracking of the mold if temperature is changed too quickly.

Fine silica dust is a health hazard. It is much more damaging to the lungs to breath than plaster. Prolonged or acute exposures to free silica can result in a condition called silicosis, which is very similar to asbestosis. This can lead to reduced lung function (read as difficulty in breathing) and possibly cancer. So a respirator is definitely in order when handling it especially when it is in a finely ground form like silica flour.

When using sand as a filler you may want to add a couple of size grades to get better packing. Its coarser texture will not necessarily lead to a coarser texture in the casting because the more liquid part of the mix will come to the surface to fill fine voids especially if used in combination with silica flour.

Crystobalite is another form of silica that is commonly used in investment formulations. It is a form of quartz with a very high melting temperature (3140°F) that has also been crushed and ground into a flour. Crystobalite differs from silica flour in that it has a higher degree of expansion. This material was discovered in the '20s and is thought to be the result in nature of lightening strikes. Now a days it is manufactured by high temperature processing of silica.

Diatomite, commonly known as diatomaceous earth, is the siliceous shell remains of microscopic single-celled algae. It is a porous crystalline material that is not as dense as silica and does not pack as tightly as silica flour. Like silica, it also becomes activated at higher temperatures and can stick to the glass. It is used in investment formulations instead of silica to form a lighter more porous mold that breaks apart easier.

Dry clay particles can also be used as a refractory filler. All clays, as described earlier, are mixtures of alumina and silica primarily differing from each other in the size and shape of their particle distributions as dictated by the processes of weathering, chemical interactions, and sedimentation that they have gone through. Some may still have unaltered fragments of feldspar or quartz. Particle size varies from 120-250 mesh for fire clays to 250-400 mesh for kaolin clays. Their addition to plaster-based investment mixes serves a refractory role by increasing high temperature strength and reducing shrinkage. Clays also act to help increase strength by providing a strong packing matrix of particles that form ceramic bonds to each other.

Molds with clay additions to binder levels require firing to 1600°F or higher to completely develop this ceramic bond through vitrification which occurs as some of the clay components melt and bond together. This temperature is a function of the purity of the clay. High percentage clay molds may be too tough for casting and will not break away easily from the glass but could be useful for cradle molds.

Non-silica refractories

Up to now we have just discussed silica-based refractories. There are also a number of other materials that can be used. Most are higher temperature refractories that will perform even better. The disadvantage is that they end up being a little more expensive than the silica refractories.

Alumina hydrate occurs naturally in a number of different crystalline forms that although they have the save chemical composition respond differently at high temperature. Dan uses it as a 50% additive to kaolin clay in making a kiln shelf separator or "wash" for his fusing work. In that application, a finely ground alumina hydrate (300 to 350 mesh) is used to give a smooth back surface to a fused piece. Coarser grinds are used for refractory additives to large open-face slumping molds that you have probably purchased. For kiln casting work, a 200-

300 mesh grade is a good choice for use as a refractory filler. Alumina does not melt by itself until about 2040°F and thus does not become activated in the range of interest for glass casting.

Zirconia is an even higher temperature refractory material that is very stable over the entire temperature region of interest in kiln casting. It does not melt until about 3700°F and it will not react with the glass. It can be purchased in a wide variety of mesh sizes from 30 to fine flours. Take advantage of this size range and add a mixture of sizes to get a high-density packing as was discussed earlier for silica.

Olivine foundry sand is made by grinding the mineral, chrysolite (magnesium iron silicate) to sand sized particles. In a fine mesh grind, it works as a good refractory additive to plaster-based investment formulations because of its low coefficient of thermal expansion. It is an expensive but extremely high melting temperature sand, 3470°F. Be aware that foundry sands, like olivine, are often characterized by an A.F.S. number that characterizes a specified mix of mesh-sized particles rather than a single mesh size. As an example, olivine 120 is a mixture of 50 to 180 mesh particles. Unlike silica sand and clays, olivine sand does not suffer from the problem of expansion at high temperature from a quartz inversion.

Property Modifiers

Property modifiers are a collection of materials formulated as part of investment mixes to change any of a number of mold properties. Among these properties are included: increased wetting of models for more faithful reproduction, decreased bubbling, increased setting time, decreased drying time, increased porosity, and internal compliance which allows absorption of the expansion or compliance of other components. Property modifiers are usually added in small proportions, not as major components, even if they are refractory in nature. They are often misused and little understood by most glass artists.

Grog is pre-fired clay that has been ground up. You will want a fine mesh grind, in the neighborhood of 120 to 200 mesh for investment formulation. It is used like a refractory to lower expansion and contraction rates of your investment mix. A similar material that is commercially available is Kitty litter. It is puffed clay and will work well for bas relief molds but is mostly put to use for making fusing slumping molds because its texture is too coarse for investment molds.

Ludo is the remains of previous investment molds that after firing and breaking them from the casting is crushed and screened. It is added to an investment mix just prior to pouring to prevent weakening the plaster matrix. Like grog it reduces mold shrinkage as well as accelerates setting time. It allows reuse of some of the more expensive refractories. You can also grind up "dead" or broken slumping molds for use as a filler. Hey no sense in wasting material. Wear a respirator when grinding this up so that you are not breathing in all the silica containing dust.

Fiberglass fibers can be added to an investment formulation to increase its low temperature strength through cure and burnout. To use them for molding, it helps to first chop them in a blender with a little water, then rinse and drain. If used dry, add a little extra water to the investment mix. If used wet, do not add the extra water. Mix them up with the water just prior to adding the dry investment.

Besides increasing tensile strength, the chopped fibers can be added to the formulation to enhance thermal shock resistance and mold porosity. They can generally be added in ratios of up to 1 cup of fiber to 10 cups of water without loosing too much detail. They will lose some of their strength near casting temperature (no surprise since they are glass after all) and become activated. But in small amounts they make an excellent modifier. Of course the fact that they are cheap and easy to come by helps some.

Addition of any fiber material reduces the ability to do extra carving on molds prior to curing. We usually chop up fiberglass insulation for this purpose, but Anna Boothe says that she has a friend who chops up the fiberglass mesh that is used in boat building or car body work into one inch lengths for her which she then uses in her investment formulations.

Alumina-silicate fibers such as Fiberfrax® can also be added to increase the mold's strength. They are lightweight, have high temperature mechanical stability, low thermal conductivity, low heat storage, and good corrosion resistance. They will not weaken at casting temperature but may become slightly activated. Use them as was described for fiberglass. They can even be used by themselves, bonded by colloidal alumina or colloidal silica, to make glass molds.

Zirconia fibers, higher-temperature fibers that will not thermally activate, are made bulk from Zircar Products Inc. They have 3-6 micron diameter fibers about 2.5 mm long of 99% cubic zirconia and can be easily distinguished from diamonds. The fibers have a melting point of 4700°F compared to the 3700°F for the alumina-silicate fibers. They show no shrinkage up to 2800°F compared to 2% shrinkage seen with the alumina-silicate fibers. Use them as described for the other fibers. They will not activate at high temperature.

Stainless steel fibers are also available to be added as modifiers to increase the strength of your molds. RIBTEC has the capability to manufacture fibers in a range of sizes from .005" to .060" in diameter (.020" is their standard diameter) and from $^3/_{16}$" to 3" in length (here the standard lengths are ¾", 1" and 1 $^3/_8$".) These fibers provide the same kinds of benefits and problems as alumina-silicate fibers. You do not need to add extra water for these fibers and you will definitely not be able to carve investment molds with this modifier added.

Kaolin clay, or **china clay,** is as was discussed earlier, a high purity residual clay. It, like all clays, is a mix of alumina and silica. When added in large amounts, it may serve as a binder or refractory. Because of its low flux content, kaolin many times does not melt until around 3300°F. Thus when added in small amounts, on the order of 3-5%, it helps in separating the glass from the mold. Its tough, durable, and easy to use. Of the various types of kaolin clay, Edgar plastic kaolin works best for this purpose.

Safe handling of investment materials

Investment materials are very dusty and at the least may serve as an irritant to your lungs. Some of the materials, such as silica flour can, as has been mentioned, cause serious damage to your lungs. For this reason, you should always wear a respirator with a filter cartridge (purple) specifically rated for silica when mixing silica bearing investments. Details for using respirators and developing respirator plans are discussed in our chapter on safety.

Mixing of investment materials should be restricted to a small designated area of your shop to avoid spreading the material all around. This area, as well as the rest of your studio, will need proper ventilation to remove any residual dust from the air. Ventilation along with other desirable hygiene practices are also discussed in greater detail in the chapter on safety.

Plaster-based investments can also be irritating to your skin. Besides for the fact that they suck the moisture out of your hands, they are mild alkalis and can produce burns similar to the lye in drain cleaners. You may find it necessary to use hand creams to replenish the moisture that is removed from your hands as you handle them. Some people seem to be more sensitive to this than others. If your skin burns, you may want to try neutralizing the alkali by washing your hands in vinegar and rinsing them well in water. Of course, the better choice would be to avoid skin contact altogether by wearing gloves.

It goes without saying that you also want to keep plaster dust out of your eyes. Wear splash goggles if you are one of those people who always seems to be working in a cloud of dust. Do not rub your eyes with your hands when working with plaster.

Investment formulations

There are a number of factors that have to be considered when you try choosing which investment formulation for use in your work. The first is based upon your glass choice — at what temperature will you be processing the glass? How fine of surface detail are you trying to capture in your work? Is your work so large that it requires very strong molds to hold it together, or so delicate that you cringe while breaking it out of the mold? Are you having problems with molds cracking during firings? Those and many other factors may affect your formulation choice or modifications you may make to it.

A number of investment formulations are presented in this section. These formulations have been obtained from various sources as well as our own personal experience and have been found to perform well in kiln casting of glass. They may be sufficient for most of your kiln casting needs. In fact, you may find the first formulation is enough. Many glass artists do not go beyond it because of its good performance and its simplicity.

With the basic knowledge that we have just presented on investment materials, you should be able to understand the purpose of the different components. Furthermore, if you are having

Investment Materials And Formulations

difficulty with an investment mix, you should now have some ideas of how to modify it to make it work better for you. When you add materials to modify a mix, you may or may not have to add extra water to the mix. The basic rule is that if the material absorbs water, add water; if not, then don't.

Remember less water generally produces a stronger mold. This is because plaster becomes a solid by the formation of tightly interlaced crystals of gypsum. As more water is added the crystals form farther away from each other and the structure is thus weakened. If your molds are fragile and crack, too much water could be the culprit. You can also try adding chopped fiber to a mix to strengthen it.

Almost all investment formulations are plaster-based mixtures with other materials added to improve either their strength or their refractory capability. As mentioned earlier, such plaster-based systems lose most of their strength in the range of 1350 to 1550°F. Mold formulations with greater than 50% plaster tend shrink unacceptably when fired over 1400°F. This causes them to crack and results in unwanted lines in the finished piece.

Mixes for higher temperatures will replace some of or most of the plaster with cement and will add various refractories to reinforce the mold. Think about these things as you examine the following example investment formulations. To further illustrate this, we will examine the effect of changing an investment's formulation as we present our first investment mix by looking at the effect of varying its plaster/refractory ratio.

Before you start mixing any of these investment formulations, you might want to review the rules that we presented on mixing plaster-based investments in the chapter on the Basic Frit Casting Process.

Investment mix # 1:

1 part gypsum plaster or cement
1 part silica flour (200-300 mesh)

This is one of the easiest investment mixes to make and it is what Dan primarily teaches. Premix both dry ingredients. The dry mix is added 2 parts of mix to 1 part of water for optimum results. Good results from this mix are also obtained by measuring dry components by volume and adding the dry mix to water until islands form on top of the water as was explained in the basic process chapter. This mix works best at lower temperatures because of the high percentage of plaster and because the silica activates at higher temperatures.

This investment formulation achieves fine detail, can be carved prior to curing, and can also be successfully added to after set. It is a soft mix and larger molds may need reinforcement with chicken wire or chopped fiber especially after the mold exceeds about 7 to 10 lb. With your current understanding of the many different investment components, you should be able to use this mix as a starting point and modify it at will to get the properties that you desire.

Frederic and Lilli Schuler did a study on the effect of varying the mix ratio of binder to silica flour for Hydrocal and Hydroperm cement on which they report in their book "Glassforming." They made a number of formulations of differing mix ratios and fired them to 1000°C. After firing the samples, they measured the percent shrinkage of each sample. Their results are shown plotted in Figure 73.

In examining the data, you can see that Hydroperm is much more stable than Hydrocal as well as being much stronger. From the data, you would think that you would want to go to as low a binder content as possible. But obviously if you go too low, there is not enough plaster to hold things together. It also seems that for ratios of plaster to silica flour above 50/50, the Hydrocal samples always seemed to crack. Below that value, although the samples didn't crack, their strength was considerably reduced.

The Schulers' recommended a 33/67 Hydroperm

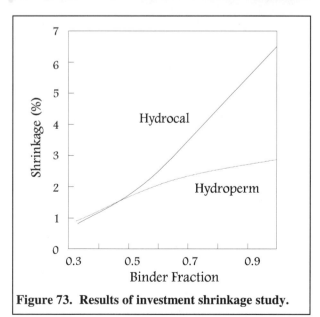

Figure 73. Results of investment shrinkage study.

mixture. You can choose which to use yourself but Hydroperm mixes are sometimes difficult to demold. You may want to add a little Hydroperm to your Hydrocal-based mix to increase mold strength.

Some artists, such as Robin Grebe, prefer to use sand instead of silica flour because of the danger of developing silicosis. Silica flour is classified as a toxic dust by OSHA, so she prefers to keep its use to a minimum. She, like many of us kiln casting artists, lives in close proximity to her studio and thus has extra incentive to keep toxic materials out. Besides some artists claim sand performs superior to silica flour in preventing cracking of the mold at high temperature because of its gradation in particle size.

Investment mix # 2:

    28   parts gypsum plaster or cement
    12   parts calcined (pre-fired) kaolin clay
    13   parts sifted asbestos
    10   parts silica flour (200-300 mesh)
    37   parts silica sand (60-80 mesh)

This is a classical investment mix developed by Rousseau in the late 1800's for his Pâte de Verre work. Since asbestos is no longer in favor, people have modified this mix in a number of ways. Some replace the asbestos with some other chopped fiber. Others have increased the calcined kaolin to about 22 parts and added a few parts of uncalcined kaolin clay. Without any chopped fiber, this mix makes a very soft mold, almost too soft for lost wax casting.

The premixed dry investment mix is added slowly to water until build up appears just below the surface of the water. Then gently stir until you reach the consistency of cream and pour. (Always wear a respirator when handling dry ingredients because all can be irritating to the respiratory tract and some like silica can cause irreversible lung damage over time.) This mold can be used up to temperatures of about 1500°F.

Investment mix # 3:

    2   parts gypsum plaster or cement
    2   parts silica flour
    1   part zircon

Add 2 parts of premixed ingredients to 1 part of water. This mix does not shrink as much as the first mix because of its higher refractory content, but at the same time this reduces the overall strength of the mold slightly. The addition of the high temperature refractory, zircon, helps reduce sticking to the glass to the mold at higher temperatures. This mix can be used for casting up to about 1600°F. It is also soft and easy to remove afterwards.

Investment mix # 4:

    4   part gypsum plaster or cement
    4   part silica flour
    1   part kaolin clay
    1-2% chopped fiber

Add 4 parts of premixed powders to 3 parts of water. After slaking, stir one of the many types of chopped fibers, prewetted into the investment and then power mix. This mold mix also achieves fine detail and works well for lost wax casting. Molds should be cured to 1450°F to mature the kaolin clay binder and are good for casting up to temperatures of around 1600°F. The addition of the clay and the chopped fiber are attempts to make the mix stronger. Even so it is still a fairly soft mix and breaks away from the glass easily. Remember that the chopped fiber will severely restrict the carvability of the mold.

Investment mix # 5:

    1   part gypsum plaster or cement
    1   part diatomite
    5%   kaolin clay

Premix dry ingredients and add 5 parts of mix to 3 parts of water. Because of the porosity of diatomite, this mix air dries faster than the previous mixes especially as wall thickness increases. This faster drying is one of the primary reasons for adding diatomite rather than silica flour in an investment mix. The porosity of the diatomite also causes this to be a lighter mix and increases its carvability. The kaolin clay, which is used here as a property modifier, helps prevent the glass from sticking to the mold allowing use up to about 1550°F. If you don't have any kaolin clay available, you can add kiln wash instead.

Investment mix # 6:

    4   parts Hydroperm cement
    1   part diatomite

Combine 5 parts premixed dry ingredients quickly with 4 parts water. Let stand for about a half of a minute and hand mix to get all the lumps out. Then

© 2000 James Kervin and Dan Fenton

Investment Materials And Formulations

vigorously mix using a power mixer about 2 inches off the container bottom until you get the desired volume increase. As some large bubbles may also be generated during mixing, you should then slow down the mixer and coax large bubbles out by raising and lowering your mixer into the mixing container.

Remember, Hydroperm has a foaming agent incorporated in it and when correctly mixed will develop a homogeneous distribution of small air bubbles that are just barely visible — about .01" in diameter. This will cause the mix to increase in volume up to ¾ of its initial volume, so plan ahead. If you vary the additives, be aware that the volume increase will vary also. Straight Hydroperm doubles in volume.

This is also a fast setting mix that can be deframed in about an hour and is ready for curing immediately. This mix has the capability to pick up fine detail and can be successfully fired up to 1600°F. One problem with Hydroperm molds is that they are fairly strong and may give you some problems in demolding your glass. So this type of mold would be used with models that have good draft and no undercuts. With careful demolding it may be reusable. Addition of about 5% kaolin clay in the dry mix will help prevent the glass from sticking to the mold. Use of a separator compound is also suggested if this is used for the casting mold. We recommend a mix like this for use in construction of a cradle mold around your casting mold.

Investment mix # 7:

- 1   parts gypsum plaster or cement
- 1   parts Hydroperm cement
- 1   part silica flour

Very similar to basic mix #1, but a little stronger and a little more prone to shrinkage. The foaming agent in Hydroperm helps to lighten the mold some allowing for faster drying and curing as in the last mix. Newy Fagan says that she uses this mix in some of her work.

Investment mix # 8:

- 1   parts Hydrocal cement
- 1   parts silica flour
- 1   part Kast-o-Lite

Premix dry ingredients. Then add by feel to water and mix by hand. This makes a strong mold that is hard to break away but is good for bas relief and cradle mold purposes. The **Kast-o-Lite** refractory from A. P. Green will make the mold a little coarse.

Alumina hydrate and alumina silicate bonded with colloidal silica also makes a very strong mold material with low thermal expansion and excellent surface reproduction capability.

Commercially available investments

If you do not want to deal with formulating your own mix, there are a number of commercial investments available that you might like to try. A number of kiln casting artists report that they have had very good success using the investments manufactured by **Ransom and Randolf (R&R)**. They say that they very rarely get any mold cracking and can usually get by without even air drying their molds.

The two formulations most commonly used for kiln casting are R&R 910 and R&R 965 investments. Both of these were formulated for casting aluminum and copper based alloys. They are both gypsum based but differ in their refractory and modifier components.

R&R 910, the stronger, denser and least expansive of the two, is reported to be formulated using calcined silica, fiberglass, and various specially graded refractories. Wetting and suspension agents are added to decrease chances of trapping bubbles on the surface of your wax replica.

R&R 965 on the other hand uses cristobalite, crystalline silica, and graded refractory additives. Both are recommended to be mixed to a consistency of 28 parts water to 100 parts of powder.

The instructions that I have been given by Donna Milliron of Arrow Springs on R&R use is as follows. To start out, she estimates or measures the volume of her mold frame. This is made easier if you use things like tin cans for mold frames. She then measures out a volume of water equal to about two thirds that of the mold frame. Into this she sifts the investment until it peaks.

To save money, you can reuse up to one part of freshly ground investment from previous molds to 2 parts of fresh investment. Here you are using the recycled material as a refractory filler and the fresh investment mix as a binder. This recycle mix is slightly weaker than the original one, but retains finer detail than the original mix. The reused material has to be kept dry just as the fresh does to get optimum performance. It also has to well ground or it will form hard lumps.

After peaking, allow about one minute for slaking and then mix. Continue mixing until the mixture will just coat your fingernail (about two minutes) and then pour your investment quickly because it will set up in a hurry. If you have problems finding this material, call Donna. She, as does Mark Abildgaard, uses R & R 965 almost exclusively in her work and may have some around for sale.

Ullmann makes a Pâte de Verre investment that is available from Ed Hoy's in 50 pound lots for about $1.00 per pound plus shipping and handling. This mix only requires addition of water and is mixed like plaster. It survives to 1500°F with very little cracking and is easily removed from the finished work. It can also be reused by grinding up a used mold and mixing 3 parts of it to 1 part of fresh investment. It shrinks a fair amount when it sets though.

Satin-Cast is another prepared investment material that gives fine detail. It is most popular with jewelers, and since they work small and use materials in small quantities, it is likely to cost a bit more than some of the other commercial mixes.

Lab plaster (they choose a real exotic name for this product), manufactured by Columbus Dental, offers fine detail, thermal shock resistance and ease of break away. It might be a bit difficult to find but the effort is worth it.

Remember there is a limit to the detail that can be captured in the glass. This limit is a result of its final strength for the detail to resist breakage during removal from the mold and during cleaning. There is also always a formulation tradeoff between ease of mold removal and durability at high temperatures. So you may need to experiment to find an investment formulation that works just right for you and your work.

Mold Construction and Processing

Now that we know more about investments, let's elaborate on how they are used in making molds. Molds are the equivalent of a photographic negative only in the three dimensional sense. It contains a negative cavity into which glass frit is consolidated to form our final positive product.

As such, the mold represents an important intermediate step between our original model and the final pâte de verre or kiln-cast piece. The complexity of its construction and the materials from which it is composed can vary widely depending upon your desires. For these reasons, we believe it is important to present as thorough an examination of mold construction and processing as possible.

This is an important subject because your final product is an exact reproduction of your mold cavity and not your original model. Thus, if you are not able to translate all that creativity that you put into making your model into a usable mold, you will not get it into your final piece. Sure you may be the recipient of a few happy accidents but the failure or success of your work depends upon being able to consistently produce a usable mold from your model.

We have already discussed in our introductory chapter on how to make single-use waste molds. These are so called because they get broken apart and wasted after being used to cast our work piece. But there is much more to learn than simply that. So let's expand our discussion on the subject to include construction of bigger and other types of molds.

Mixing large batches of investment

Earlier we discussed how to hand mix small batches of investment. As your work gets larger, your batches will have to get larger and hand mixing may no longer be appropriate. With large batches of investment, you may want the assistance of power tools with mixing attachments similar to an eggbeater, which by the way can be used for medium-sized batches.

Attachments designed for paint mixing are ideal for this purpose. They can be used with a variable speed power drill to mix your investment. Mixing attachments come in many different sizes and shapes with many looking like two or three bladed airplane propellers or outboard motors. They may be mounted singly or multiply to a shaft.

One of the best power drill attachments available for mixing investment is illustrated in Figure 74. It is called a Jiffy mixer, and is available from your neighborhood paint or hardware store in a variety of sizes. The two vertical blades help confine the mixing action within the mixer while the horizontally opposed impellers force material from the lower and upper areas of the container into the mixer. The leading edge of the lower impeller can scrape the bottom of the container bringing all the heavy particles into the mix while the ring prevents puncturing the container walls.

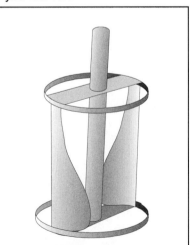

Figure 74. Jiffy mixer used in mixing large investment batches.

© 2000 James Kervin and Dan Fenton

The design of the Jiffy Mixer also helps minimize splashing and sucking of air into the investment mix as long as you keep the mixer head well below the surface of the investment. They are available in a wide variety of sizes suitable for mixing of investment batches anywhere from a pint to 100 gallons.

One thing to be aware of when choosing to power mix investment is the possibility of electric shock. Electricity and water don't mix too well. We know that you will try not to spill any water but sometimes in the heat of mixing investment we all get a little sloppy.

Figure 75. A simple vacuum cleaner deaerating setup

To avoid getting shocked, you should consider using doubly insulated drills with good ground wire connectors. The use of extension cords with Ground-Fault Interrupt (GFI) protection built in, as most new homes have in bathrooms, is also a good idea. A GFI will cut off the power to the drill if you get a short and hopefully protect you from receiving too great a shock.

Deaerating investment

If you are lucky enough to have a vacuum pump, it helps at this point to pull a vacuum over the mix for about a minute to try and remove any gas bubbles from the freshly mixed investment. Otherwise they can end up becoming attached to your model and will show up as glass lumps on your final piece.

A simple low-power vacuum system can be realized using a rubber-gasketed lucite plate with a wet-dry shop vacuum hose connected to it. To use it, you push down on the top of your mixing bowl with the vacuum on to pull air out as shown in Figure 75. The sides of your bowl should be fairly high in order to prevent accidentally vacuuming up the investment mixture because some inevitably will boil up. This kind of accident can be hard on your vacuum.

Another way to remove the bubbles that get into a mix is to vibrate it. The easiest way to this is just to bounce the mixing bowl on the table for a little bit. A more elaborate system that works a little better is to hold a vibrator against the side of the bowl. Later you can use it to work out all those aches and pains that you got from crushing your own frit or whatever else ails you. There are also small flat vibrators that jewelers use or if you are working really large scale a vibro-lap may do the job.

Alternate mold frame construction methods

We have already discussed how to construct a relatively simple mold frame box from cardboard and hot glue. Another convenient thing to have around from which you can rapidly construct round molds are strips of linoleum. Remnants can usually be obtained cheaply, if not for free, from a flooring contractor. Good sizes to prepare ahead of time are strips varying in width between 2 to 8 inches and ranging in length from 8 to 24 inches. They are rolled into cylinders and taped together using duct or plastic packing tape.

You can also use heavy roofing paper or light-gauge sheet metal, if you have any of that lying around. When needed either of these strips can be coiled into a cylinder and held in place by duct tape or clothesline. With roofing paper, you will have to wrap it 3 or 4 times around to make a strong enough mold frame.

The Tokyo Art Institute has advocated the use of semi-rigid acetate sheet in their book and it is another good frame material. They suggest sheet material between 3 to 5 mils thick (0.08 to 0.13 mm) because it is rigid enough to hold its shape but not so rigid that it can not be rolled into a cylinder. Use it like linoleum. You can also cut up old place mats

Figure 76. Coddle made from linoleum strip.

Mold Construction And Processing

if you have them. Acetate has the advantage in that being fairly thin, you will not get as large of a step discontinuity on the side of your mold where the material wraps around itself.

After you have rolled up the cylinder for any of these alternatives, the edges are then hot glued down to your pallet or clay dams can be built around them. A mold frame constructed in this way is commonly referred to as a coddle (good to know for those scrabble freaks out there).

Early mold makers used leather strips for this purpose. These were tanned and waterproofed using soap. They would usually keep a number of widths available and would use one by wrapping it around a short round base. They would tie it in place with rope and would patch any cracks between the leather and the base with clay.

Coddles are usually formed into cylinders because this is the shape into which the hydrostatic pressure of the fluid investment will try to force it. But since coddles are constructed from flexible materials, they can also be formed into other shapes such as ovals or rounded triangles. Unfortunately because of the flexible nature of these materials, non-cylindrical shaped coddles will require some reinforcement to hold their shape.

If you find yourself getting seriously into the production of kiln cast glass, you may want to make some convenient, adjustable wood mold frames. Adjustable wood mold frames should be made from closed grain hardwoods like maple, cherry or birch. These are usually preferred over softwoods like fir and pine because they absorb less water from the plaster and warp less. In addition, it helps to seal the wood with something like polyurethane.

The adjustable frame illustrated in Figure 78 could be constructed from four 1" x 4" boards about a foot and a quarter long. Attached to the end of each board is a three-inch right angle bracket that fits down over the next board. Since each of the brackets is only screwed into one end of the boards, the boards can be slid relative to each other to relatively quickly construct sturdy rectangular or square mold frames 4" high of

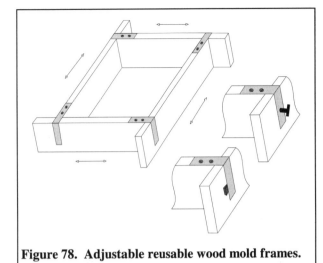

Figure 78. Adjustable reusable wood mold frames.

up to about 1 foot on a side.

To help seal the boards against each other, you can insert small wedges underneath the angle iron or add T-bolts to push them together as shown in the figure. If you need deeper frames, you can use 1" x 6" or 1" x 8" boards along with correspondingly larger angle brackets. If you need bigger mold frames, you can use longer boards.

The only problem with these fixtures is that you are restricted to rectangular frames, which may have excess mold material in the corners that is not needed. These areas can be blocked off with cardboard or clay as desired. The frames are held down on the glass pallet by weights placed atop of the wood frame or by clamping them to the table. If you encounter leaking problems, pack some clay around the base. Hot glue can also be used, but it may take a wood chisel and some time to remove the glue.

You can also make irregular mold frames using sheet metal. This cones in handy when your models are very irregular shaped and you do what different sections of your molds to have a lot of thickness variation in them. This is the trick that Mark Abildgaard uses when he casts his spirits as seen in Figure 77.

If you find that your production work is consistently small and about the same size, you may want to consider using a

Figure 77. Using irregular shapes mold frames.
Photo by: Mark Abildgaard

jeweler's-like flask. This is useful for items like beads. To do this, find a regular metal can, like those used to package vegetables, that is about an inch or two larger in diameter than the largest dimension of your models. The can should not be made of aluminum or zinc. Aluminum will melt. Zinc is not compatible with glass and its fumes can have serious health effects.

Cut off the bottom and top lids. If still too long, trim it such that it is about ½ to ¾ of an inch longer than the model mounted on its sprue and reservoir. Make a small base of clay around the reservoir and push your flask can into it. You are now ready to cast your mold into the flask.

After the mold is setup, you really don't have to remove the flask if you don't want to. In fact you may find its strength helpful to alleviate problems associated with mold cracking. You can then proceed on as normal with drying, curing and casting. Just be sure to keep the flask away from the coils in the kiln and go a little bit slower in curing because water will not be able to get out the sides of the mold with the flask in place.

Reinforcing molds for frit casting.

A common problem with molds is cracking during curing or casting. This can be caused by a number of factors. The first, which has already been discussed, is that of plaster investment material that was not stored properly and was allowed to hydrate. The second is incomplete drying of the mold such that there was too much moisture left in it. This moisture then breaks the mold apart as it turns into steam during the early portion of the cure cycle. This can be partially alleviated by going through this portion of the cure cycle slower, but this can be hard to manage if you do not have a controller on your kiln.

Another cause of cracking is improper screeding of the bottom of your mold such that it does not sit flat on the kiln shelf. This can be corrected by lightly sanding the bottom of the mold flat after it has set up and before it is cured. Sandpaper works best if the investment is relatively dry. A "Sure Form" tool by Stanley, as shown in Figure 79, also works well for this process if used on the mold before it is completely dry. Remember to avoid inhaling the dust.

Figure 79. Using Sure-Form tool on a mold.

We have also discussed how firing a mold causes the mold material to weaken and how large molds may crack from outward pressure of the molten glass.

Well what can we do to correct these problems? Some can be handled by better process control during screeding, curing and casting. Alternatively you can strengthen or reinforce your molds. We have already discussed how to strengthen a mold by using stronger binders like cements or clays as well as strengthening modifiers like ceramic or metal fibers.

Another internal strengthener that you could use would be to embed a metal mesh like chicken wire into the outer surface of the mold. Cut up and fit your chicken wire so that it leaves a gap between it and the model to be filled with investment. If not it can end up discoloring your casting an ugly brown color. Embedding wire mesh into a mold can also make it harder to break apart after casting, possibly resulting in damage to the casting. So be careful.

For really large molds, you can use a stronger wire mesh like that used for stucco or concrete. This is what Linda Ethier uses for some of her really large castings. She suggests that you wire the corners together for increased strength.

Another reinforcement option, that of providing external reinforcement for your mold, may be a better option. One way to do this that we have already discussed is the use of a jeweler's-like flask with your mold. This provides great strength but ends up being size limiting and requiring longer curing.

Another simple technique that is often used to strengthen molds is to wrap the mold with 18 to 20-gauge wire. Stainless steel wire is best for this purpose because of its strength and heat resistance but it is hard to find. We get ours from Truebite.

Because of the greater availability, most artists tend to use copper wire.

Wires are usually applied in horizontal bands about the mold at about one-inch spacing. They are tightened by twisting them with pliers at two opposite corners of a mold as shown in Figure 80

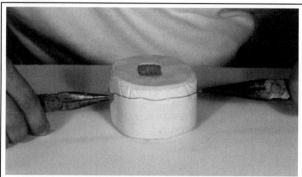

Figure 80. Tightening wire around a mold.

The wires can be applied before or after curing. Dan usually applies the wire before curing when the mold is fairly strong and this is what we recommend if you are using stainless steel wire. If using copper wire, apply it after curing and then, only to molds with cracks because the copper wire gets badly oxidized as well as softer at the high temperatures present in the kiln.

If you decide to apply wire post cure, go slow with the pliers when twisting the ends of the wire because you can easily crack the mold, which is actually quite fragile at this point. This would ruin your whole day. Hopefully the wire will hold the mold together so that those hairline cracks do not grow into gaping fissures.

Another way to provide external strength for your mold, which we suggest that you consider, is casting a cradle mold about your casting mold after it has set up and the mold frame is removed. Add a second frame about the casting mold to allow for an additional ½ to 1 inch of thickness on the bottom and the sides of the casting mold. Then cast the cradle mold in the second frame about your casting mold using an investment mix such as #6 or a high temperature castable refractory as well as the adding hardware mesh.

Cradle molds provide a strong mold around the casting mold while still allowing the mold directly around the casting to be broken away fairly easily. If you make the cradle mold without a bottom, it may be reusable and could be used as a mold frame in which to cast your next mold. The disadvantage of a cradle mold is the extra thermal resistance that it provides. You will have to fire your casting just a little slower to account for this.

Stacking hard firebricks or some other strong refractory material like slices of kiln shelves around your mold can also approximate a cradle mold. Use your imagination to come up with other ideas. For large molds, another good idea is to construct a cradle structure out of ceramic fiberboard that you wire together. Any of these cradle alternatives will provide the best support possible if you cast your investment molds in them.

Multi part molds

Most of the molds that you will make and we have described are single-part waste molds. We concentrate on these because we are into making one-of-a-kind pieces of art. There are times though, when you may want to reuse a mold or are making a mold around a model that can not be burned out.

Removal of the model might be possible with a one-part mold if your model has no undercuts and sufficient taper to it that the finished piece can be removed from the mold without destroying it. If not, you are forced to construct a mold with more than one part to achieve reusability.

There are also situations where you want to intricately decorate a piece but it is impossible to reach into a single-part mold that will give the right shape. Such molds require more steps than a single-part mold but are not fundamentally any more difficult to make.

Adding a reservoir section

An example of a simple second part of a mold would be to add a frit reservoir to a mold. Many times when you start out on a model you may not be sure how big it will be and thus how big a reservoir that you will need to fill it. If you make your initial mold and measure the volume of the modeling material, you can then construct a reservoir of the correct size. In fact, a lot of times you can construct the reservoir model from the same material that you took out of the mold. In those cases, you will be sure to have a reservoir large enough to handle up to fifty-percent shrinkage.

When making a reservoir, try to get the bottom opening to be about the same size as the opening into the mold or just slightly smaller. That way you will not get material hung up in the reservoir nor trap air at the joint under it.

Attaching the reservoir section to the rest of the mold is usually done by slathering about a quarter inch thick layer of investment around the joint between the two pieces. Of course, the usual caution on dampening the two sections of the mold before adding new investment still applies. No additional reinforcement is usually needed because the glass does not really exert much force on the joint. This is because the opening sizes are relatively the same and there should be relatively little material left in the reservoir.

If the opening of the reservoir into the mold were small relative to the opening of the mold and there was a relatively tall column of glass in the reservoir, it would be possible that the molten glass could try to lift the reservoir component off the mold and break the joint between them. In those cases, you might want to add wire staples across the joint like we discuss soon in the mold repair section.

Basic two part molds

The simplest multi-part mold is a two-part mold. To make such a mold possible, the model must have a shape such that you can run a line around it, where on either side of that line the model has no undercuts and good draft. When you plan your two-part mold, look over your original model. Try to determine how you will have to construct the mold so that it allows disassembly.

As a first example of a two-part mold let's look at making a model of an open hand. In this situation, the obvious way that the mold should made to be disassembled from the model is in the direction away from the palm and the back of the hand as seen in Figure 81 and not in the direction of the fingers or the wrist. If you look at your hand you can see how if you were to make a mold without undercuts that the mold will have to separate down the middle of you fingers.

If your hand is flat then this will be pretty much a flat parting surface. If instead your hand is cupped

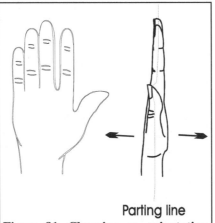

Figure 81. Choosing an orientation for making a two-part plaster mold.

slightly, the parting surface will no longer be a plane. We shall use such a cupped hand to explain how you can go about constructing a two-part mold.

The first thing that you have to identify is the parting surface for the mold. Do this by taking the hand original and laying it down on your glass pallet. Position the model such that the nominal separation directions are up and down. You may have to prop the original up in one direction or another by using some clay shims.

Next determine the parting line by rubbing pencil lead on a small carpenters square and then running the square around the object transferring lead to the high points of the object. If you have problems with this, just use a marking pen to mark the points the square touches as seen in Figure 82. This line should extend all the way around your object when you are finished and represents the line at which the mold should separate.

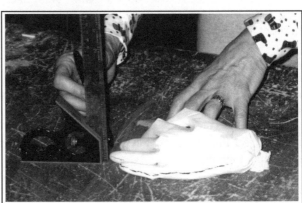

Figure 82. Determining the mold parting lines.

Once you have determined the parting line, block off the lower half of the hand on your glass pallet using clay as shown in Figure 83. The clay should come all the way up to your parting line. Construct your mold frame around the hand. Extend the clay bed out to the mold frame to fill its lower half. Smooth the clay out to make a uniform parting surface.

Add half of a sprue to the model for filling during casting by building up some of the clay on one end of the model. Spray on a parting agent of your choice. Then invest the upper half of your mold with plaster-based investment using the standard mixing procedures and allow it to set up.

Mold Construction And Processing 91

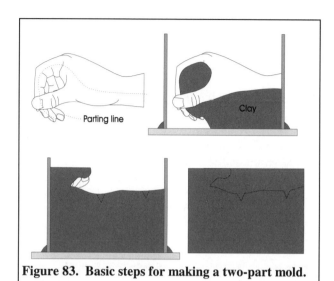

Figure 83. Basic steps for making a two-part mold.

Remove the mold frame and clean out the clay from the bottom of your original. Since it only takes about an hour for plaster to set, the clay should still be moist enough to be reusable. Carve some natches, features to align the pieces when they are put together. Many artists use V-shaped natches because they are easy to carve into the investment. Others carve little spherical hollows into the first half of the mold's mating surface using melon balling tools or a knife that has had its tip bent in the shape of a U. Some artists believe that this type of natch is superior to sharp natches because they weaken the mold less.

In either case, you are always left with the dilemma of how deep to make your natches. The deeper they are the more vulnerable they are to damage. We find features one-quarter to one-half inch to be sufficient.

Invert the mold with parting surface up as in Figure 83. Add the other half of the clay sprue. Construct a new mold frame about the bottom half of the plaster mold. Here is where the premade wood frames come in handy. You can just slip them together with very little effort. If needed, moisten the first section of the mold, spray the rest of the hand and the plaster with your parting agent. You have to remember that step or you may not be able to separate the two sections of the mold. Make the release agent layer thin because sometimes they can interfere with the cure of investment.

Then invest the upper half of your mold using standard practices. Once set up, your mold should now be complete. You should be able to open it up and remove the hand and the sprue. This mold is now ready for detailed filling. After the outer layers are decorated with a layer of glass paste, the two halves are put back together and secured.

Securing mold parts together

One thing that has to be considered when constructing multi-part casting molds is developing a way to hold the parts together during the casting process. Some artists put flanges on the parts that they can clamp together using a piece of metal that is bent in the shape of a U and acts like a paper clip holding the parts together. Others hold the parts together by wrapping them with wire. You can also use steel hose clamps to hold them together.

Others use the technique shown here of casting a cradle mold around the multi-part mold for the final casting step or you can make it so that the multi-part mold slides out of the cradle mold. To do this you would cast your mold parts so that the assembled inner mold before casting the cradle mold has a cross section of a trapezoid with the large base being the reservoir opening. This arrangement is shown in Figure 84. You can see how the draft on the multi-part mold allows it to be easily slipped out of the cradle mold.

If you do not want to go through the trouble of building a second coddle for the cradle mold, you could just slather investment on the outside of the joints of the two-part mold instead. This will not be quite as strong as a cradle mold that you cast in a coddle but it should work.

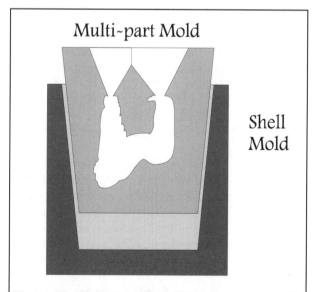

Figure 84. Using a cradle mold to capture a multi-part mold.

© 2000 James Kervin and Dan Fenton

To slather on a layer of investment, put your mold top down on your pallet. Glass is again an excellent choice for this because it will not soak up water from the investment. Mix up some more strong investment as we have done before but let it start to set up some. When it starts to get to the consistency of whipped cream, spoon some out and slather it onto the outside of the two-part mold.

Shape and form the investment layer with a spatula. Build up at least ½" of thickness for strength. Try to get it all spread out before the investment sets up too much. If you do not, the layers will not bond well to each other.

Any of the reinforcement techniques that we have discussed can also be combined together. The most popular combination seems to be to wrap the mold with wire and then to slather some more investment over it. These two techniques work well together because the wire rigidly holds the pieces together while the slathered investment helps protect the wire from the effects of heat.

Press molds

Another commonly used type of two-part molds is a pressing mold. This type of mold is used to make large flat objects like plates or shallow bowls. For a model with much height, it would be hard for the glass to flow down the narrow interface between the halves of the mold and it may not fill completely. This flow would also tend to wipe out any decoration that you might try to apply onto either surface of the plate.

So instead, what we do is make the two halves of the mold. Apply any decorative features to the surface of the plate using the techniques that will be discussed in the next chapter. Next build up a layer of frit about 50% thicker that the final thickness of the plate on one surface of the mold. Then place the other half of the mold on top of the frit.

The mold will be held open by the extra thickness of the frit. As everything heats up and the frit gets soft, the mold will start to close. Adding a weight onto the top half of the pressing mold will help to press the two halves together. This will also help to reduce motion of decorations during consolidation by forcing the two halves together and capturing the frit.

So let's look at how to do make and use a pressing mold in a little greater detail. When making a pressing mold, it is easier to construct the concave section of the mold first. Before the days of political correctness, this was usually known as the female half of the mold. It is the half that shapes the bottom of the plate or the bowl.

Start by modeling the bottom of the plate or bowl out of clay. The bottom will have a ring or feet to stand on and some curvature to it. If you are making a bowl with a lot of curvature to it, you really need to decide whether the right type of mold is a pressing or a regular two-part mold. On either you will get some pattern motion, but a real steep-sided pressing mold may also allow too much glass to flow out of the mold before it closes. This results in an incompletely filled casting.

Figure 85. Setting up mold frame to cast bottom half of press mold.

Build up the model of the underside of the object on a glass pallet. Clay is probably the right material for this task although turning of a plaster shape may also be a good option because it allows you to develop a uniform model.

You are not worried about the how thick the plate or bowl will be at this point but you do need to add any external decorations to it. Build an oval shaped mold frame the model. The frame should be about the usual one half-inch thickness at the top and two opposite ends but it should be about three times that thickness on the two opposite sides of the mold ninety degrees from the other two sides. The two thicker sides serve as locations in which to carve mortises.

Mortises are female cavities of an assembly joint. You may be familiar with the term from furniture. There the mortises are the cavities carved into the wood into which the male half of the joint, the tenon, is glued. For our situation, they are there to guide the two halves of the mold together when they close

Mold Construction And Processing 93

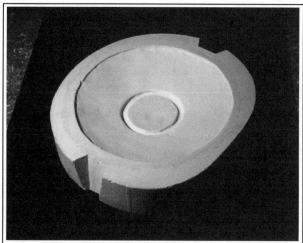

Figure 87. Construction of a pressing mold mortise.

in the kiln during our pâte de verre casting run. See Figure 87 to see what a mortise looks like.

So cast the concave half of the mold around the model of the bottom of the plate. Allow it set up like a normal single part mold. Then remove the concave mold from the glass pallet and clean the model out of the inside of the mold.

Now carve the mortises into the thick ends of the mold. We find the ideal tool for this job is a box knife that uses ordinary safety blades and a sharp wood chisel. We use the razor blade of the box knife to cut the two outer sides and the bottom of the mortise and the chisel to carve the inside. The investment cuts are easiest if you use both tools as you would a chisel, pushing the blade straight into the piece, rather than trying to saw into them. Notice how the surfaces of the mortise have draft both to the top surface as well as the outside of the mold. These surfaces should be smooth so that the tenon will not hang up in the mortise.

The depth or length of the mortises depends upon the thickness and the curvature of the plate. They have to be engaged throughout the casting process. The reason that the thickness matters should be obvious. It is because the layer of frit will be about 50% thicker than the final thickness of the plate. The curvature is important because the greater the curvature the higher the male or convex half of the mold will be sitting out of the female mold when assembled. This is because the convex portion will be hitting out near the edge of the mold before the bottom and thus are held up higher.

Anyway, we usually cut our mortises to be at least three times as long the intended thickness of the plate or up to six times longer if we are making a fairly steep sided bowl. The thickness is usually twice the mold thickness at the top and a single thickness at the bottom. In this case they would taper from about an inch thick at the parting surface to about one-half inch thick at the base of the mold.

Your mortises do not have to be shaped exactly like those in the figure. Sometimes the shape of your mold may suggest a different shape. Just make sure that they are long enough and have sufficient draft to allow separation of the two mold parts after casting them.

Next roll out a slab of clay to the thickness of your plate. If your plate is going to have a variable thickness, usually thicker in the center and thinner at the edges, roll out a slab of clay the thickness of the thinnest section of the plate. Build up an extra thick area in the center of the mold with another smaller clay sheet first and then lay the full-sized sheet over it. We find this to be much easier than trying to roll a variable thickness slab or trying to thin down a constant thickness slab once it is installed in the mold. It also ends up giving us a much smoother surface.

Trim the rim of the plate model. Add any sculptural components to the top surface of the plate. Adjust surface texture as desired. In this case, Jim first added some clay rocks and some squishy lizards from Natural Wonders™ to the top of the plate. You could also add wax or lampworked additions to the plate.

Figure 86. Preparing to cast top half of press mold.

Next we cast the top or convex half of the pressing mold. We start by coating the top surface of the bottom part of the mold and the mortise cutouts with release agent to allow disassembly of the pressing mold. Use a little oil or liquid detergent for this. Try

© 2000 James Kervin and Dan Fenton

not to use too much because some release agents can interfere with the setting of some investments. You could also lightly coat the surface of the clay if you want but remember that you will lose some detail from the model. If the bottom or concave half of the mold is not fresh and damp, you may want to dampen it slightly first so that it will not pull a lot of moisture out of the second half during casting.

Construct a coddle tightly around the bottom half of the mold such that it covers the whole outside surface of the mold and about one inch over the top of the mold. You are making the convex half of the pressing mold a little thicker than normal to give more strength to the tenons. If you are using a thick material to make your coddle you might want to start the wrap of the coddle wall at the same location as in the first casting to minimize the investment that seeps down along the side of the coddle. Otherwise you can stuff a little clay down along the top edge of the concave mold piece to contain the investment.

If you have any appreciable gap down along the inside of your coddle, you might want to attach it to a pallet with a small ring of clay. Now mix up enough investment to fill the coddle and gently pour it into the coddle trying, as usual, not to trap any air bubbles.

Figure 88. Separating two press mold halves.

After the mold has cured, remove the coddle and split the two halves apart. Do this by locating the joint between the two halves and gently driving a spatula it on the opposite sides 90° from the mortises and tenons as in Figure 88. You might also want to start off with driving a safety razor blade in around the joint. A crack should eventually appear at the joint. You may have to tap the spatula in around a little bit to get the crack to go all the way around the mold. Be careful near the tenons and do not tap too hard or you can break the mold. We like to have the mold sitting flat on the surface of our worktables when we separate the halves so that we do not lose control of them.

Once the crack extends pretty much all the way around the mold, stop tapping. Now gently try to pry the two halves apart. Try to lift straight up to avoid cracking the tenons. If you should happen to damage a tenon, repair it as will be discussed at the end of this chapter.

After you have the two halves separated, removed the clay and clean the molds. Wash the two halves out thoroughly with water to get out all of the clay residue. Carve any extra details that you want to put into the plate. Place the two halves together and cut some registry marks into one side of the mold. These will be used to remember which side of the mold is which. They will also be handy in helping align the two halves as you put them together. This can sometimes be a little tricky as the half of the mold on top floats around on all the frit.

You will also need to carve some gates on the flat mating surface of one of the halves to allow excess glass to flow out of the mold. There is always some excess glass in our castings because we usually slightly overcharge the mold to avoid having voids in them. Carving gates allows more of the excess glass to flow out the gates and the pressing mold to more completely close.

The gates are usually carved into the concave section of the mold because you can carve all the way across the face without having to worry about wear to stop. We cut at least four radial gates approximately equally spaced and located as shown in Figure 89. These gates are carved after both halves are cast so that they will not get replicated in its mate.

Figure 89. Carving gates to allow for flow of excess glass in press mold.

Mold Construction And Processing 95

Figure 90. Two halves of press mold finished and ready for application of glass.

Form the clay that you have removed from the mold into a rectangular solid and calculate its volume. This is the product of the length times the width times the height. When you cast the plate, charge the mold with about one and a half times this amount of frit by volume. If you want to be more exact, use the density technique (explained in next chapter) to calculate the weight of frit that you need to add to the mold.

Now paint in any raised decorative areas using the techniques of the next chapter. Allow decorative areas to dry before proceeding. Your glue holding in the decorations will have to be good and strong if the decoration is on the convex side because this side is usually placed on top during the firing and will thus have to be turned upside down during assembly. For fairly shallow curvatures on a semi flat plate this may not be absolutely necessary and the convex half could be put down.

Fill the bottom half of the mold with as much of the frit as you can, trying to contain the frit within the model area of the mold. Again the concave half of the mold is usually on the bottom. We typically try to put all the frit to fill the mold in at this time since we can calculate approximately how much glass is needed. Try to fill all the way to the edge of the plate and slightly shape the pile to match the shape of the convex half of the mold, which gets placed on next. You may even want to use the upper half of the press mold to shape the frit pile before adding raised frit features

Next fit the other half of the mold down over the pile aligning the two halves using the registration marks on the side and the mortises and tenons. Try get the gap between the two halves to be equal all the way around the mold as in Figure 91. Try to have the frit pile touch all the way to the edge of the top surface of the model. Try not to rub the molds on the pile as you do this so as not to disturb the surface decorations. Instead lift the top half off and sweep some of the frit from the large gap side over to the narrow gap side.

If you did not get all the frit into the mold before assembly, try to put the rest in at this point by spreading it equally around the inside of the gap. Use the eraser end of a pencil to tamp this glass into the cutout section of the mold without disturbing the gap distribution. This can also be done to any glass that fell off the pile and into the mating surface area during assembly of the two halves. You want to keep those surfaces clean as much as possible so that the mold will close properly.

Satisfying this requirement and filling to the edge of the model section on the top half of the mold can be hard to satisfy together. This can be aided by using a little binder around the edge of the frit pile.

Now fire the mold. Use weights on top of the upper piece of the pressing mold to help force them together during the firing. Use steel weights of between two to four pounds. Depending upon the size of the mold, you will place the weights in the center of the mold or distribute them equally over the surface of the mold.

With a pressing mold you can be pretty sure that some glass will spill out of the mold and onto the kiln shelf. How much depends upon how much you over charge the molds. So first of all make sure that you kiln shelf has a good coating of kiln wash on it. Next sprinkle a little more or some plaster, whiting or sand around the outside of the mold to catch some of the extra glass and prevent it from flowing over the edge of you kiln shelf. Also spread some around on the bottom of your kiln.

After firing, you again separate the two halves of the molds by placing them on your worktable and tapping at the upper half of the joint above any glass

Figure 91. Press mold assembled and ready to fire.

© 2000 James Kervin and Dan Fenton

flashing that may have flowed out of the mating surface. Try not to break any of this flashing off as it can cause cracks to flow back into the bulk of the piece. Gently break the mold off the glass, again trying to direct any blows onto the plaster and not onto the glass.

After all the investment has been removed. Finish the plate as described in the chapter on finishing. Trim any excess glass off the edges of the plate with a diamond saw either using successive straight cuts with a trim saw or going all the way around with a band saw. Polish it with polishing equipment and sandpaper.

Multi-part molds

Now we are going to get a little more complicated. Back in the hand example, if your hand had been flexed differently, it might not be possible to construct a two-part mold that could be removed from the hand. You might need a three or four-part mold. In that situation, you would start almost identically as with a two-part mold. The difficult part is trying to determine where the parting lines should be. This means that you have to take time to study the original model. Think about which surfaces have the proper requirements to serve as a mold section—no undercuts and good draft. Then go and mark your parting line on the model.

Make a bed of clay on some cardboard as before. Choose which section of the model you want to cast first. Block the model in with clay such that only that section of the model is exposed. Build up the mold frame, apply mold release to the clay and the model, then cast the first portion of your multi-part mold and cut natches as before.

Repeat procedure for each adjoining mold segment until all sections of the mold have been cast. Separate the mold sections and remove all traces of the model and the clay separators. Paint in any details with glass paste, reassemble the mold and secure the sections together as discussed earlier. Fill the rest of the model cavity and reservoir with the remainder of your frit and fire away.

The more parts that your mold has the more likely they are to leak. For more information on making multi-part molds see either *Plaster Mold and Model Making* by Chaney and Skee, *The Complete Book of Pottery Making* by Kenny or *Mold Making for Ceramics* by Frith.

Hollow vessel molds

One example of the use of multi part molds is for making hollow vessels. Another is statuary pieces. Both of these applications require controlling access to the surface of the piece to get the final shape, thus inhibiting the ability of being able to decorate the surface of the object. Hollow vessels require at least a two-part mold configuration because you need a plug to form the hollow. They look similar to pressing molds except they get completely assembled and then filled through a sprue.

Figure 92 illustrates the components of a hollow vessel mold used for single color or poured layers of color. The terminology is the same as that used in foundry work. The portion that shapes the inside of the vessel is called the drag perhaps because you have to drag it out after the casting is completed. The part that forms the outside is called the cope. Both of these pieces can, if desired, be captured in a shell mold. The shell mold may or may not allow disassembly of the cope and the drag.

Such a hollow vessel mold is cast in a couple of steps. Figure 93 illustrates the basic steps for making a single-pour hollow vessel mold. The way that we go about this is to first establish the general shape of the vessel by having an insert of the general size and shape that we want for the inner surface of the vessel. A lot of time this will consist of a bowl, vase, or drinking glass.

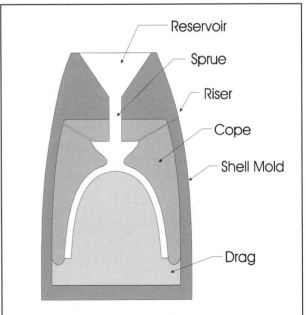

Figure 92. Hollow vessel mold similar to those for metal casting.

Mark Abildgaard

"Ancestor Boat" 1996
Bonded kiln cast glass
Size: 28" x 28" x 9"
Photo by: the artist

"Winged Figure" 1999
Bonded kiln cast glass
Size: 21" x 11" x 4"
Photo by: the artist

Anna Boothe

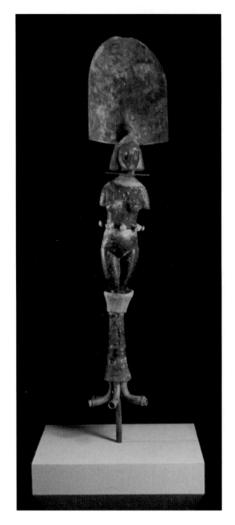

"Raspberry Bouffant Bowl" 1996
Pâte de Verre
Size: 10" x 7½"
Photo by: Eric Mitchell

"NASA Notwithstanding" 1993
Lost wax kiln cast glass
   with copper and steel
Size: 36" x 8" x 4"
Photo by: Eric Mitchell

"Neo-Flower Bowl" 1996
Pâte de Verre
Size: 8" x 8¼" top x 3½" bottom
Photo by: Eric Mitchell

Ruth Brockmann

"Group of Coyote Figures" 1994
Fused glass with stone
Size: 9" to 12"
Photo by: the artist

"In Memory of Walter B." 1996
Fuse cast and slumped glass
Size: 14" x 14" x 4"
Photo by: the artist

"Turtle Dance" 1998
Fuse cast and slumped glass
Size: 16½" x 4"
Photo by: the artist

Linda Ethier

"Ancient Athletes" 1992
Kiln cast glass and neon
Size: 60" x 84 x" 36"
Photo by: Michael Mathers

"Entering Eden" 1999
Kiln cast glass with inclusions
Size: 18" x 9" x 3.5"
Photo by: Bill Bachhuber

Newy Fagan

"Cactus #1" 1992
Kiln cast glass
Size: 13" x 9½" x 1"
Photo by: Mike Barrett

"Horse Head #2" 1988
Kiln cast glass
Size: 8½" x 1½"
Photo by: Mike Barret

Mary Fox

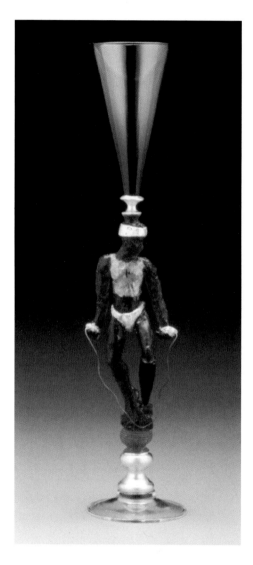

"The Determined Veteran" 1996
Lost wax kiln cast and blown glass
  with gold and copper
Size: 18¾" x 4" x 4"
Photo by: Bill Bachhuber

"Beauty Rules" 1998
Lost wax kiln cast and blown glass
  with gold
Size: 17½" x 4" x 4"
Photo by: Bill Bachhuber

Robin Grebe

"Mind's Eye" 1994
Kiln cast glass and metal
Size: 32" x 15" x 10"
Photo by: Will Howcroft

"Building Tradition" 1994
Kiln cast glass, metal and slate
Size: 37" x 10" x 11"
Photo by: Will Howcroft

Rachel Josepher Gaspers

"Twist Series #1" 1986
Kiln cast glass
Size: 12" x 6" x 2"
Photo by: the artist

"Ammon's Hair" 1996
Kiln cast glass & concrete
Size: 25" x 20" x 8"
Photo by: the artist

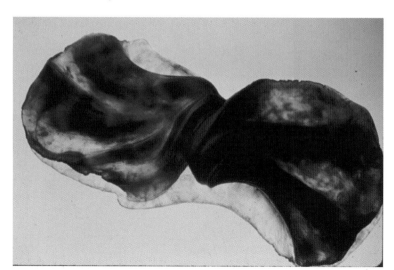

"Twist Series #3" 1986
Kiln cast glass
Size: 10" x 7" x 2"
Photo by: the artist

Lucartha Kohler

"The Language of the Goddess"
Kiln cast glass with slumped &
 sand carved plate glass
Size: 17" x 12" x 20"
Photo by: the artist 1994

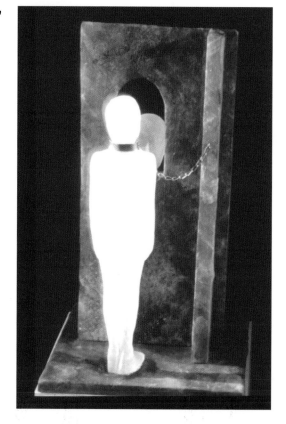

"Love Story III" 1994
Kiln cast glass with brass
Size: 20" x 12" x 12"
Photo by: the artist

Donna Milliron

"Cityscape, Whisper and Dreamscape Beads" 1994
Pâte de Verre / Cire Perdue
Photo by: Christopher Marchetti

Floral Vessel
Pâte de Verre
Size: 7" x 2¾"
Photo by: the artist

Copper/Glass Vessel
Pâte de Verre with copper
Size: 6¼" x 3"
Photo by: the artist

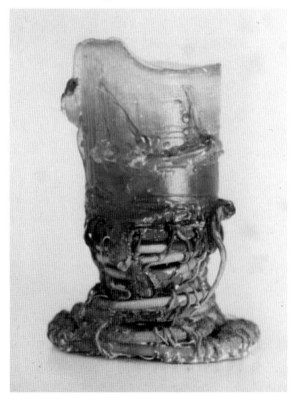

Charles Miner

"The Ladies" 1996
Kiln cast glass
Size: 13" x 13"
Photo by: Addison Doty

"Tarpon School" 1996
Kiln cast glass
Size: 13½" x 18½"
Photo by: Carol Wright

Seth Randal

"Double Cage Cup" 1994
Pâte de Cristal
Size: 19" x 9"
Photo by: Roger Schreiber
Collection of Antonio Amado

"Amphore Classique D' Albrâtre"
Pâte de Cristal 1993
Size: 23" x 103"
Photo by: Roger Schreiber
Collection of Mr. & Mrs. Marvin Weis

Alice Rogan-Nelson

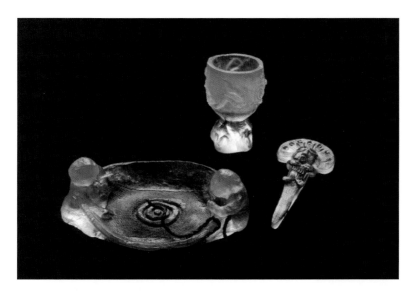

"Mythical Setting" 1995
Cire Perdue kiln cast crystal
 with copper and silver
Size: knife   3½" x 6½" x 2"
       goblet 5¼" x 4" x 4"
       plate   4" x 9½" x 6½"
Photo by: Richard Nelson

"Celestial Explorers" 1995
Pâte de Verre
Size: 9" x 18" x 4"
Photo by Richard Nelson
Collection of Michael P. Curry

David Ruth

"Nebula" 1995
Kiln cast glass
Size: 49" x 40" x 11"
Photo by: the artist

"Baltra" 1996
Kiln cast glass
Size: 16½" x 10" x 9"
Photo by: the artist

Janusz Walentynowicz

"Seagull" 1992
Lost wax kiln cast glass
Size: 29" x 22" x 19"

"Carousal" 1993
Lost wax kiln cast glass
Size: 30" x 18" x 15"

Mary Frances Wawrytko

"Diva" 1994 (Edition of 8)
Cire Perdue kiln cast glass
Size: 7" x 9" x 4"
Photo by: the artist

"Edris" 1995 (Edition of 8)
Pâte de Verre
Size: 9" x 8"
Photo by: the artist

Mold Construction And Processing

We coat this core shape with a release agent like an oil or soap and build up our vessel model around it. Nothing says that this piece could not be a turned shape of investment and thus end up as our drag. We then develop the basic thickness of the vessel using our modeling media of choice—wax or clay. Although from the diagram, it should be obvious that our model design is of wax because it would be too hard to dig out some of the base. Upon completion of this, we build up any external surface decoration and add a base, a sprue and any other gating that may be required. The mold frame is then built up around the piece and the cope is cast.

Note how we have added air vents coming off from the bottom of the base. These vent models are made from wax rods that we extruded. They will prevent trapping air around the perimeter of the base as the mold fills. These can also be used during the firing to monitor how deep the heat has penetrated into the mold. When you see glass coming out of the vents, you can be pretty sure that you have a good fill.

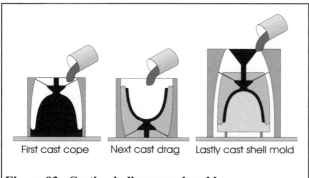

Figure 93. Casting hollow vessel molds.

Next we invert the mold, remove both the bottom of the mold frame, and the central core former in order to set up for pouring of the drag. Here is where you find out if you applied enough release agent on the shape model. If you want, you can also decorate the inner surface of the vessel at this point by adding material to or removing it from the inside.

It is probably also apparent that it would be nice to use the same mold frame (except for the bottom, which you had to remove) for this investment pour also. This will not be possible if you made your mold frame from cardboard, which looses its strength after it gets wet. Coddles made from linoleum work just fine for this. If you use the same mold frame, you can probably get away without putting a bottom underneath it, but we do not suggest it because the new investment will loose moisture too quickly through it.

Remember to carve some natches into the exposed surface of the cope so that it will be easy to align the cope and the drag later if you are going to disassemble them. Before casting the drag, rewet the exposed surface of the cope and apply a parting agent to it. Here a slightly more viscous agent like Vaseline is preferred because it will help minimize water loss to the cope and interpenetration of the investment.

At this point, you might also want to allow for some air vents between the mating surface of the drag and the cope by laying some wax rods in radially a few places around what will be the lip of the vessel. They will help allow the lip to fill. These should be very tiny though or you can have problems with them breaking off a section of the rim of the vessel. Try to keep them under the size of a matchstick or definitely no thicker than one half of the vessel wall thickness at the point of contact. Of course if you are going to disassemble the two halves of the mold, you can carve air vents into the mating surface of the mold parts after you disassemble them.

For extra strength, you might want to cast a shell mold around both the cope and the drag. If you do, you can cast it, in which case you will need to build up another mold frame around it, or you can slather it on freehand. Smooth out the surface of the cope and drag as well as ensure that they have sufficient draft to be removable from the shell mold. Add any required extensions to the gating system such as reservoir systems or air vents and of course some more of the viscous parting agent before putting it in place in the mold frame. You may want to add some internal reinforcement to the shell mold by placing it inside the mold frame at this point. Now mix up a strong investment formulation and pour the shell.

Another way to make multi-part molds for hollow

Figure 94. Mold for casting hollow vessels.

vessels was passed on to us by Anna Boothe. She says that she learned the technique from Karla Trinkley. In this technique, she uses a multi-part ring or step mold to build up a vessel with lots of detail. As an example, the mold for her "Neo-flower" bowl is shown in Figure 94.

Here she has a solid core drag and a multi-part cope. To hold them in register, she carves sharp natches into them as she casts each successive layer. The layers are constructed so that the bottom of the bowl remains exposed. Casting in this orientation allows the walls to fill better and makes them less likely to crumble or slump during the casting process. By casting the mold in steps, she can build up the frit in shorter sections, which she can easily reach into to decorate and then adding each successive layer after the previous one gets filled.

This kind of mold is still cast very similarly to the process that was illustrated in Figure 93 except that the cope is cast in multiple pieces and she did not make a shell mold. To hold the whole thing together during casting, she forms a cradle mold by wrapping the exterior with wire mesh and lathering investment over it.

For hollow vessels, instead of building multi-part molds, the inside of the vessel could be packed with pre-fired mold material and chopped fiber paper. This takes the place of the drag as will be seen in the next chapter. This mixture is pressed gently but firmly against the inside of the cavity to hold the glass frit into place during the final firing. If done properly this technique can prevent sliding of the glass in the vessel walls during the final firing. A solid cylindrical drag could be run down the center of the fiber paper if desired.

Mold construction without frames

As you have probably guessed by now, you do not always have to construct a mold frame to make a mold. This goes for both single-part and multi-part molds. Constant cross section molds cast in mold frames can have vastly varying wall thickness to them. This causes variations in the mold's thermal resistance that can lead to temperature variations in the casting during the cooling phase of the firing cycle. If you do not compensate for this by slowing your cooling rate, you could end up with cracked castings.

To get around this problem, you can construct molds without mold frames to get a more uniform mold wall thickness. This is done using some of the same skills that we discussed in making plaster models. In this example, we will discuss making a two-part mold.

In this demonstration, we will discuss the use of metal shims in dividing a clay model into sections for casting of the mold. These shims are trapezoidal pieces. Since we make molds about ½ to 1 inch thick we use shims about 1½ high. They will be about in inch wide at the base and one-quarter inch at the top. We stick them about ¼ to ½ inches deep into the clay model.

They are cut from thin strips of tin, brass, or aluminum as seen in Figure 95. Any of these three materials can be purchased in rolls or sheets about one and a half inches wide. Aluminum is often preferred because it does not react with the water in the investment to discolor the mold as brass does. But then aluminum shims are also not as strong as brass shims, so you will need to decide which kind you prefer. We like tin best because they are stronger and do not seem to interact with the investment.

Figure 95. Cutting metal shims for mold casting.

To use the shims, start by determining the high point of your model along which separation of the mold should occur. Then place the shims to divide the model up into two sections. Do this by sticking the small end of the trapezoid into the model to a depth of about a half-inch. Start at the bottom of the model and stick the first shim in so that one edge is along the surface of the pallet so that you get a good edge on the mold here.

Stick the next shim, again small end into the model, right next to the first so that they overlap each other some. Try not to have gaps show between the shims. This goes both between the edges and the flat surfaces.

Mold Construction And Processing

Figure 97. Placement of shims into a model.

To keep the parting line smoother, each shim will have the two shims on either edge side of it on the same surface side—either the front or the back. To further smooth the mold parting line that the shims will create, you could lay tape over the side of the shims that you will cast first. Also, as you insert the shims into the model, you want to keep the amount sticking out of the model approximately constant as in Figure 97. This will help you judge the thickness of the mold as you form it. Drawing a line on your pallet the same distance from the model will help also.

Next seal the surface of the model with a couple of coats of acrylic spray paint. Depending on whether your pallet is of a porous material or not, you may also want to coat a three-inch or so section around the model. Not only will the coating help to keep the clay from penetrating into the mold during the investment process but it will help keep the model from getting messed up in our next step.

Since working with investment freehand can be messy, we next cover up the back of the model. For this, we use dampened newspaper. It should be just wet enough so that it just sticks to the model. Build up a layer made up of small pieces on the back of the model until it covers the whole thing.

Next apply a coat of release agent on the front of the model. You can paint, dab or spray it on. Just don't let too much build up or let it form any froth.

Now you are ready to apply the investment to the front half of the model. Mix up a batch of your favorite investment. After slaking, stir up the investment. Continue to stir it until it starts to thicken. When it gets to about the consistency of heavy cream and will just start to peak, it is ready to apply.

As when pouring the first investment layer, you want to make sure that you do not trap any air against the model. You can put this layer on with a brush or your fingers. The traditional way to apply it is by the backhanded flick method. This will throw the investment against the model with just enough force to drive it into any recesses on the surface.

To execute the flick, grasp a small amount of investment between the tips of your fingers and your thumb. With the back of your hand facing the model, give an upward flick of your hand extending your fingers toward the model as they come up. You will end the motion with palm of your hand facing the floor and your fingers pointing toward the model as illustrated in Figure 96.

Figure 96. Executing the backhand flick.

If this is your first time doing this, don't be surprised if it does not go exactly where you want it. It takes some practice to develop control. Don't bother to try and clean up yet. Get the investment on the model first. Start from the top of the model and work your way down. Use gravity to your advantage. As the first layer flows down the face of the model coax it into any hard to reach places by blowing hard on it. Don't blow too hard though or you might just blow the investment off the model.

Often an inexperienced mold maker will start putting the investment on too early before it is thick enough. If this happens, don't worry about it. Just scrape it up and try again. As the investment thickens, build up extra thickness on the model until is about an even three quarters of an inch all the way around the front surface of the model. About a quarter of an inch of the separator wall of shims will still be showing.

Smooth up the investment coating. Many times doing this by hand will give you a good feeling for

© 2000 James Kervin and Dan Fenton

how thick the investment layer is. Add some extra investment at the top to form what will later be the base of the mold during casting. Make it big enough that the mold will be stable. Use a sheet of glass to form a flat for the base on the top of the mold as seen in Figure 98. Again use the position of the shim walls to know how thick the base is. Now let the first layer set up.

Let the first half of the mold set up for at least an hour. Then remove the shims to expose the parting surface of the mold. Smooth up this surface as desired and the surface of the model where you pulled the shims out. Carve at least four spherical natches into surface of the mold for registration. Carve one on either side of the base, one at two o'clock, and one at ten o'clock.

Clean any loose investment off of the parting surface of the mold and off of the pallet. Slowly fold back the newspaper at the edges on the backside starting at the top. Trap any loose investment in the newspaper as you fold it back until you have the whole back surface of the model exposed. Draw your line at one inch on the pallet around the back of the model for judging the thickness of the investment.

Next apply a coating of release agent both onto the back of the model as well as onto the parting surface of the mold. Now apply investment with the backhand flick like before to form the back half of the two-part mold. Remember to keep air bubbles off the model and to make the investment coating a uniform three quarters of an inch thick. Add extra investment again at the top for the base and flatten it again with a sheet of glass.

Let the mold set up for at least an hour. Tap all the way around the model at the joint between the two halves. As it starts to open, gently pry the two halves apart trying to keep the gap uniform all the way around the parting surface as you go. The suction between the mold and the model will resist your efforts. As the gap gets bigger you will be able to use your fingers to help pry them apart. Go slow so as not to damage the mold.

When the two mold halves have separated, dig all the clay out of them and clean them up. Wash out any clay residue. If you are just going to cast a single-colored solid casting, then reassemble the two halves and wire them together.

Next we have to construct the reservoir. Use a piece of paper to trace around the mold where the bottom of the model used to be. Then invert the mold, reposition the paper on the same way and trace the shape of the opening into the mold. This piece of paper will help us size the opening of the reservoir into the mold.

Take the clay that you just removed from the mold and shape it into a cone or pyramid that is flat on top. In Figure 98 we actually reconfigured the model to form the reservoir cavity connection using the two feet. Shape the top to match the tracing of the hole into the mold.

Figure 98. Forming the reservoir model.

Now mix up some more investment and invest it around the cone leaving the top uncovered. Flatten the top of the reservoir mold with a sheet of glass and try to shape it to match the outside tracing of the freehand mold. Then allow the reservoir mold to set up.

When the mold has set up, clean the clay out the reservoir volume. Dampen all three sections of the freehand mold and assemble them with the reservoir

Figure 99. Sections of a freehand mold ready for assembly.

up. Mix up some more investment and slather it over the outside wires and joints not more than a quarter inch thick. If you have doubts on how stable things will be during casting, then build up the base some. When everything has set the mold is ready for kiln casting.

What happened

Sometimes we think that we have done everything correct but we ended up with mold voids, burrs, or cracks. We know it could not have been anything we did wrong because we never make mistakes. Unfortunately that is not true. We do make mistakes and luckily the hard-learned lessons last the longest. Let's discuss some of the things that can go wrong in mold making and what we can do to fix them.

The problems that we encounter in mold making generally fall in to four main categories:
- problems with the materials
- problems during investment
- improper mold design
- mold handling problems

Problems with the materials refer to improper storage or preparation of the investment. Problems during investment refers to mistakes we make as we are casting the mold. Improper mold designs refers to problems in judgment of how to design a mold. Mold handling problems refers to things like dropping or overheating a mold. Let's look at each of these general areas and discuss some of the things that can go wrong.

Problems with materials

One common problem is that our investment may be old. We discussed earlier how investment can pick up moisture out of the air. You should store it in waterproof containers or put the bag of investment inside a plastic bag.

The next common mistake is mixing investment in improper proportions with water. Either too little or too much water can prevent developing the proper interlacing set of crystals to give the investment strength.

Improper techniques for mixing the investment with the water can also cause problems. Always add the investment to the water and not vice versa. Do not try to use a batch of investment that has already started to set up. Even if you try to mix it up, add more water or extra investment, you will not get good results. Pour your investment faster or make a new batch.

Problems during investment

The most common problems encountered during investment occur from incomplete fill around the model. This usually is a result of trapping air bubbles against the model during the pour. Paint that first layer on, add wetting agents to the model, do not pour directly onto the model, pour slower, and tap or vibrate the mold frame to release any air bubbles. All of these will help minimize trapping of air bubbles

Larger incomplete fills are usually due to other factors. Improper orientation of a model is a big one. Do not have recessed areas that because of their orientation can trap air. Make sure that the investment is well mixed and not lumpy.

Tears in investment can occur if the investment is disturbed after it has started to set. Screed the exposed surface of the mold while it is still fluid. Then set the mold frame aside out of the way and let it set up. Don't rush it.

Break up of the mold can occur if the mold not made in a single pour. Try to correctly calculate how much investment you need to make a mold. If you have to mix extra investment in order to have enough. And if you have to add additional investment to a mold let it set up first and make sure that you dampen the mold before adding more.

Multi-part molds can break in disassembly. This could be from trying to rush things as you take them apart, but it can also be due to failing to apply a release agent to the surface of one part as you cast the next.

Improper mold design

One common mistake is to let one or more wall of the mold get too thin. Many times this can be on the base, where we did not make the mold frame high enough. Thin sidewalls can happen if the mold frame is not rigid enough and it flexes under the weight of the investment. Another cause might be if you did not firmly fix the model to the pallet and it moves around on you.

Another common design flaw is putting the seams of multi-part molds at inappropriate locations. Try to chose locations of low detail so you can remove flashing or burrs. Pick points of strength not of weakness.

Natches or tenons can break off if they are designed too thin or without sufficient draft. Round melon-ball natches are less subject to problems than triangular ones.

Mold handling problems

Molds are delicate. They have to be handled carefully to avoid damage. The most common damage that we see occurs during cleaning a model out of a mold. During this process, the mold can get scratched or gouged. Take it easy and go slowly.

The next most common mistake occurs in disassembling multi-part molds, especially pressing molds. Ensure that you get the split to extend all the way around the mold before you try to separate the pieces. Separate them gently and evenly. Do not cock them.

Multi-part molds should be stored assembled and secured together until you are ready for casting. Besides the ease with which tenons and matches can get damaged, you can also damage the flat mating surfaces. The pieces are tied together so that they do not fall apart when picked up.

Next, as we will discuss shortly, you need to be careful when drying or curing a mold so as not to thermal shock it. Heat it at the recommended rates.

Lastly, there is normal handling damage. Try not to drop or in any other way to physically abuse a mold.

Repairing molds

There will be times in mold construction when not everything will go right and your mold will need rework. Maybe one of the previously mentioned problems happened or maybe it was something else entirely. The first thing to assess is what went wrong.

If this was a problem with the material such as old or poorly mixed investment, the whole mold may be weak and fragile. Thus even though you fix the immediate problem, further problems may pop up as you work with the mold. It may crack again during casting and create a big mess in your kiln. In this case, it might be better to chuck the thing and start over. That is of course unless you put a lot of time into making the model and did not make a master mold to replicate it. Then you might want to try to save it.

If your mold was damaged through mishandling or just has minor flaws in it, then it is usually worth the effort to fix it. Let's discuss how you might repair some of the different things that can go wrong with a mold.

Filling in mold bubbles or voids

For incomplete fills or air bubbles trapped on the surface of a model, you have to add more investment to fill in the voids and resculpt it. Before adding more investment to a mold, you need to moisten it so that it will not steal moisture out of the investment that you are adding. This would result in the new investment being a lot weaker because it did not developing the proper set of interconnecting crystals. We just dunk the mold in a bucket of water and drain it out for this.

Now mix up a small amount of investment. Mix it up a little wetter than usual because no matter how well you dampened the damaged mold it will still steal water from the mold. Apply the new investment with a damp brush or spatula. Shape it as close as possible to the desired features of the model. Use whatever tools necessary. Our favorites are fingers. Allow the investment to set up. After it is set up, carve any missing features into it. Allow the mold to dry and then touch it up with a little sandpaper to smooth out any transitions from the mold to the patch.

Removing burrs or flashing from a mold

Sometimes your mold frame will leak and extra material or burrs hanging off of a mold will develop that you have to remove. When casting multi-part molds, investment may wrap around one of the earlier cast pieces. In either of these cases, you will have to trim the extra material off of the mold using a knife or scraper. We usually dampen the area that we work on because we feel this makes it a little easier to work and less likely to fracture.

When trimming investment off of a mold, it is much better to make a number of shallow cuts than a single deeper one. That way if the investment fractures it will not break off too much of the mold. You will also be less likely to get unwanted breaks if you work from corners inward to the center of a side and at the edge of the flaw removing material little by little.

Mold Construction And Processing 103

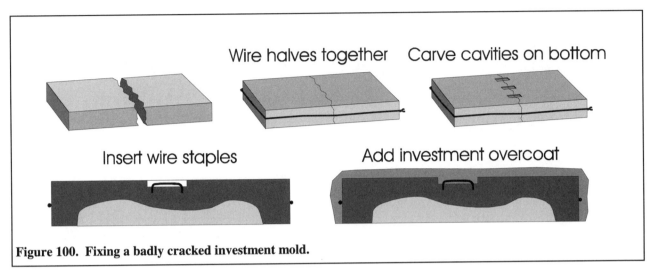

Figure 100. Fixing a badly cracked investment mold.

If you do get unwanted damage as you are trimming an edge, you can go back and fill the area in with investment as we discussed for air bubbles.

Repairing cracked molds

There are a couple of ways to fix a cracked mold. You might want to apply a bandage to hold together the cracked parts or you might want to recast a portion of the damaged mold.

Now when we suggest bandaging a damaged mold, you have to understand that do not mean putting a Band-Aid on it, giving it a little kiss, and sending it on its way. We mean bandaging it with a little wire and adding investment to hold it together.

Let's say that you had a mold cracked in half as is depicted in Figure 100. You want to hold these two parts together during the kiln casting process. The first thing that you want to do is handle the parts carefully. We do not want to induce more damage in the parts. If they have completely separated, gently separate them and blow off the mating surfaces to remove any loose pieces. This will allow them to fit back together better.

Next fit the pieces back together. Take some wire and bind the pieces together by wrapping it around the pieces. Stainless steel wire is best if you can find it because it resists the heat. Nichrome wire is also a good choice and may be easier to find but is likely to be more expensive. We recommend you do not use copper because it is too soft and will not resist the heat. It can get completely oxidized and crack with a single firing.

Tighten it up with a twist at the ends of the wire and half way around the mold. The second twist, which is started first, will give better control over how well you can tighten the wire. If you are working on a rectangular mold, it works best to position the twists at the corners of the mold.

Even though this wire is holding the pieces together this is not enough. At temperature, the wire will expand more than the mold because it has a larger COE and will get loose. This problem is worsened the larger the mold and thus the longer the wire that wraps around it has to be. So we need something more than this.

We need to attach short wire "staples" across the crack to hold it together. Since these staples are short, the effect of the relative growth differences between the wire and the mold is minimal. Make the staples at least an inch long and be sure to get the corners nice and square. Clip the tines of the staple so that they do not extend more than half way through the thickness of the mold at the point of application.

Next tap the staple lightly into the surface of the mold. We are not trying to drive it into the mold because this could further crack the mold. Instead we are just trying to mark the location where the staple will penetrate the mold. Now drill pilot holes just slightly smaller than the wire diameter at those locations. These holes will help prevent cracking of the mold as you install the staples. Again be careful not to drill too far into the mold. Now push the staple into place, tapping gently if necessary. Apply these about ½" to ¾" apart across the bottom and on the sides of the mold.

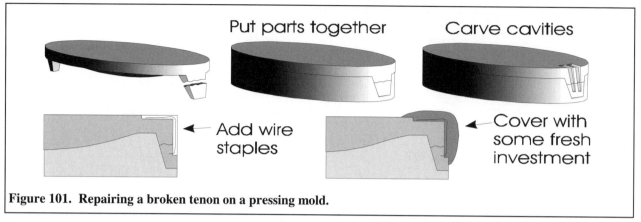

Figure 101. Repairing a broken tenon on a pressing mold.

Next cover the staples and the wire wrapping around the mold with a light coating of investment at least one quarter of an inch thick. This will help protect the wire from heat damage. We suggest that you coat the whole bottom of the mold and level it with a piece of glass so that you do not point load of the area of the crack. As always when adding extra plaster to the mold dampen the mold thoroughly first. If you do not get the bottom of the mold completely flat, level it in a bed of sand in the kiln when you cast in it.

After you have finished with the exterior of the mold, turn it over and see if you have any patching to do on the inside of the mold. Fill any missing material and shape it as we suggested for air bubbles.

Replacing pieces cracked off a mold

Another common mold injury is to break all or part of a tenon off of a pressing mold. To fix this, we suggest making two slightly different staples like those seen in Figure 101.

The first thing that you do is bend up your staples as shown in the figure. Again clip the length of the top tines so that they will not be longer than half the thickness of the mold at this point. These staples will help hold the new section of tenon that we are going to cast next in place. Carve shallow grooves down the side of the part for the staples to lie in. This will help make them more stable and the joint stronger.

Lift the top section of the pressing mold off and apply a thin coating of parting agent to the mortise and the surrounding area of the mold exterior. Replace the damaged top on the bottom half. Place one staple in the grove and check that the bottom tine will not stick in too far. Tap the top of the staple to mark the location. Drill your pilot hole and tap it into place. Do the same with the other staple.

Now dampen the mold pieces and mix up a new batch of investment. Use it to coat the staples at least one quarter of an inch thick and to fill in any missing section of the tenon. Try as much as possible to keep the investment off the exterior of the bottom half of the mold. Let the investment set.

Gently ply and lift the two halves of the mold apart. Carefully trim excess flashing off the edges of the tenon.

Salvaging poorly mixed molds

About the only thing that you can do with a mold cast from old or poorly mixed investment is to cast a cradle mold around it to hold it together. Place it opening down on your pallet and apply a slight rim of clay around it. This will prevent the new investment from getting underneath the mold and into your model cavity. Next form a mold frame to cast at least a one half-inch thick cradle mold around the sides and the exposed bottom of the mold. Cut and form wire mesh reinforcement to fit into this cavity. Then cast the cradle mold around the mold using one of our stronger investment formulations. After the investment sets, remove it from the mold frame. Remove the clay and you are ready to go.

Wax melt out

Casting artists that use clay will dig it out and rinse the mold clean before proceeding with the glass casting. But for the many artists that use wax as their main modeling material, the first step after investing the mold is to melt or steam the wax out of it. Which method you use depends upon what equipment you have available and how much of it you do. Steaming seems to do a better job than just

Mold Construction And Processing

Figure 102. Mold set up in the kiln for wax melt out.

melting wax out but we all have the basic equipment required to melt out wax—a kiln.

Melting wax out in a kiln

Ancillary melt out equipment to go with your kiln for wax melt out is cheap and simple. You need a pan to hold water and a rack to hold the mold over the pan.

For a pan, use a cheap metal baking pan about three inches deep. We feel square or rectangular shaped ones are best for our work since many of our molds end up being shaped this way. This pan should be dedicated for the purpose of melt out and never used for cooking again. This is because wax is hard to clean out and not necessarily good to eat.

You could purchase a rack to hold the mold over the water. Many baking pans have a matching rack to hold roasts out of the juices as they cook and these racks would work just as well for melt out. But you could also make your own. Quarter-inch mesh hardware cloth works well for this purpose. For best support, cut a piece about 2 inches longer and wider than the top of the pan. Then cut one-inch squares out of the corners and bend over the flaps to strengthen the edges of the mesh. This will help prevent the mesh from sagging under the weight of the mold.

To set up for melt out in a kiln, place the pan in the kiln and put the grill in place. Pour about a half-inch of water into the pan. Then place the mold, sprue cup end of the mold down, on the grill so that the wax will drain into the water as shown in Figure 102. You should have removed any clay from the mold ahead of time because it won't melt and will block the wax.

Make sure that the water is not touching the bottom of the mold as this can also impede wax flow as wax floats on water. At the same time, do not let the mold get too far from the boiling water or you may not get the mold hot enough to melt the wax. It also works better if the mold is wet because that will help prevent the wax from wicking into the mold.

Now you are ready to cook. Melting the wax out of a mold does not necessarily have to be done in your kiln but this is usually the most convenient heat source available in the studio. We recommend not doing it in your oven. Turn on the kiln and set it to soak at about 400°F. The trick here is in making sure that you don't run out of water. The water, besides for providing some steam to help melt out the wax, also prevents the melted wax's temperature from getting any higher than the boiling point of water. Otherwise the wax can burn and possibly start a fire.

Although, we have been told by Mary Francis Wawrytko and others that they just catch the wax in a pan and never bother with the water without any problems, we do not recommend it. If you do this it is suggested that you keep the kiln temperature fairly low on the order of 225°F so that the wax will not get hot enough to burn.

If you run out of water in the pan when melting wax out in your kiln, the best thing to do is to refill the pan in place. We know that it sounds a bit bizarre to pour water into a hot kiln but its really a lot safer than pulling the pans filled with molten wax out of the kiln and refilling them outside the kiln. Spilling hot wax on yourself is not exactly what the doctor ordered. Just make sure that the kiln is turned off and cooled back down below the boiling point of water before adding more water.

If you have concerns about spilling water all over the inside of you kiln, you could use one of the long oil funnels available from an auto parts store for the job. It will allow filling without getting too close. A metal funnel is a better choice for this operation if you can find one.

During the melt out run, keep an eye on the kiln to make sure that it doesn't climb over 450°F. If you have a controller, this is not a problem. Also make sure to vent the top of the kiln with a small bone. When the surface of the water gets covered with wax, go into the kiln and examine some of the molds. Use a good set of gloves and remember to turn the kiln off first.

© 2000 James Kervin and Dan Fenton

Figure 103. Canning steamer set up to dewax molds.

Make sure that you melt as much of the wax as possible out of your mold or you will have that much more smoky material coming out of the mold later during the burnout and heat-curing phase. If you can get most of the wax out, you may be able to forgo a separate curing and burnout run. Once you decide that melt-out is complete, remove the mold from the kiln.

Be sure that you also remove the grill off the pan since it is may be more difficult to remove once the wax hardens. If using a kiln, it's a safe idea to let the wax cool in place for about a half-hour before trying to remove it. The wax will cool faster if the pan is on kiln furniture. Once the wax hardens, you will have some sheet wax, which you can use in making your next model if you decide to reuse wax. If done right your molds will come out of the kiln clean and free of wax residue.

Steaming wax out

As we stated earlier, steaming wax out of a mold seems to work much better than melting it out both in how much wax it removes and how clean it gets the molds. In steaming you also do not load up the bricks in your kiln with a lot of moisture, which is not necessarily good for them. Steaming also seems to go much faster because you are not waiting for your kiln to heat up and cool down. So with so much going for it, we prefer removing our wax this way.

An ideal piece of equipment for steaming wax out of open-faced mold is a canning steamer. It is real convenient to use in your studio by heating it on a hot plate. It has a nice heavy rack to hold the molds over the lower pan filled with water and a high top to put over the molds. Figure 103 shows a run set up in the bottom half of a canning steamer by Donna Milliron.

If you don't have a canning steamer, you can easily make a simple substitute by using heavy wire mesh to construct a rack that you stick into a large pot with a lid. Then you just add your mold and a little water and soon you will have a nice pot of wax soup. Doesn't that sound just delicious. The trick here, just as in melting wax out in a kiln, is to keep the pot filled with a little water or you might end up with blackened casting molds and a visit from the fire department.

Steaming wax out of an open faced mold this way will usually get it clean enough that you will not have to go through a burnout phase later. If you cannot afford a canning steamer, you can do as some artists we have heard of and construct a large steamer by cutting old 55-gallon drums up into sections.

For large molds other than open face molds, steam injection seems to work better then steaming. Steam injection, if done right, can also get your mold clean enough that you will not have to go through burnout. To be able to steam-injection clean your molds, you need a system to generate the steam and channel it into your mold. For this, you need an industrial steam pot to first generate the steam to which you attach some heavy black rubber tubing with a nozzle on the end which can be constructed out of copper tubing.

Figure 104. Using a steam injection to remove wax from a mold.

Figure 104 shows Anna Boothe's setup in her studio. She sets the molds up on an old bed frame with the mold openings pointed down and injects the steam from underneath. She runs her steam system at about 15 psi. The trick is to feed the steam through the nozzle into the mold opening and ensure that you melt all the wax out. Be sure to wear heavy gloves and splash goggles when you do

Mold Construction And Processing

this for protection from the steam and the dripping hot wax.

A poor man's steaming system can be constructed from a pressure cooker. The ideal thing is to remove the fitting from the top and attach a tee into it so that the base of the tee points to the side and the top goes up and down. The original fitting is then attached to the top section of the tee pointing up and the hose is attached to the base pointing to the side.

You might be wondering why we just didn't attach the hose to the top fitting and use it like that. Well, if the end of the hose got plugged by sticking it into wax or investment, the pressure cooker could build up enough pressure to blow its lid. Believe us, you don't want that to happen. Although we have heard of people doing exactly this and getting away with it. Some people will try anything.

We have also heard of people using wallpaper steamers and even vaporizers to steam wax out of molds. Look around, you may figure out a new way with something that you already have around your home. You could probably figure out how to get a teakettle to work with a little ingenuity.

The last method for melting wax out of a mold is to nuke it in the microwave. A long-time mold maker, Billy Hiebert, gave this idea to us. He says that he has had good success with it on molds up to six inches in diameter and nine inches tall. A few molds that may have had too much trapped water did blow their bottoms off but most worked just fine.

You can not get around this by air drying the mold ahead of time, because if you do, nothing will happen. An interesting thing to note is that a microwave does not really melt wax. They are tuned to the vibrational frequency of the hydrogen-oxygen bond in water. Therefore you could microwave a dry mold all day and nothing will happen. Saturate your mold with water first. This will provide steam to melt the wax out.

If you want to use this technique, you are limited some from size unless you have a big industrial sized microwave. Set up your mold in a glass baking or a plastic microwaving dish with the reservoir down and the mold spaced up from the bottom of the dish enough to let the wax flow out. Then just punch up about 25 to 30 minutes on high and away you go. Of course, we would not recommend doing this in a microwave that you are going to use for food again just on general principles.

Drying investment molds

After melting out and cleaning your mold the next step is to dry and cure it prior to use. This will help reduce the risk of it cracking during the casting run. As described earlier, there are three stages to curing plaster-based investment molds. The first, which will be discussed in detail here, is air curing or drying. The other two stages, low-heat curing and mid-heat curing to remove physically bound and chemically-bound water, will be detailed in the chapter on kiln procedures.

Air-drying allows the binder in an investment, usually plaster, to thoroughly set up. Drying will take anywhere from several hours to several days depending upon the drying conditions and the mass of the mold. Conditions, which enhance drying, are low humidity, increased airflow, and higher ambient temperature.

If you want to accelerate the process, you can construct a chamber to enhance air drying like that illustrated in Figure 41. It is made using a cardboard box and a portable droplight. To make it, first cut a small 1" x 4" vent in the top of one side of the box and a similar one at the bottom of the opposite side of the box. Put the mold under the top vent and the drop light near the vent in the other side. This will warm up the inside of the box and will set up an air circulation pattern from the bottom vent to the top vent to help remove any moisture released by the mold.

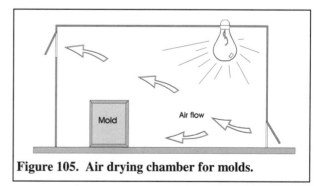

Figure 105. Air drying chamber for molds.

An alternate heating method would be to forget the light and the lower vent and instead poke a hole through that side of the box for a cheap hair drier. Push the hair drier into the hole and turn it on low so that it will not burn itself out. Old hair dryers are another one of those items that can always be found cheap at a flea market and usually only the high speed might have been lost.

The larger the mold, the longer it will take to air dry. This is because water vapor has to work its way through the torturous path between the plaster crystals to the surface of the mold. Drying time ends up being an exponential function of the thickness or size of the mold, such that doubling the size of the mold may increase drying time by much more than a factor of two. Incomplete drying of a mold can cause it to blow apart during heat curing or casting.

So take that extra time to air dry the mold, it is well worth it.

You can increase a mold's ability to dry by adding foaming agents to your investment like detergents. Diatomaceous earth or materials that burn out like sawdust will also help to increase a mold's porosity.

Artists trying to achieve precise placement of color through the use of pastes will want to apply the pastes to a damp mold to prevent the paste from drying too fast. Any air-drying is then done after application of the paste is complete.

Advanced Casting Techniques

In this chapter, we discuss a few advanced kiln casting topics. This will include how to achieve variations in translucency, color, and texture. Gaining mastery of these techniques can add real pizzazz to your work. To that end, we will discuss the making of glass pastes and their controlled placement in a mold. This technique will add to the salability and uniqueness of your work. So its mastery is important.

Controlling the look of your work

Different effects can be achieved in your work by varying how you mix and fill your molds with glass frit. These techniques vary in complexity from the very simple to the very detailed. The simplest glass placement technique is to fill your mold with a homogeneous load of frit. Here addition of the "frit de jour" will result in full-density, single-colored pieces.

These single-colored pieces can be reinvested and cast in conjunction with single colored pieces of other colors or glued together to create larger multicolored pieces. This process is useful when you are trying to make multicolored pieces where you want to minimize any blending of colors as the frit slumps during melt and fill of the mold.

Even with one homogenous colored frit, a number of different effects are achievable. The first of these has to do with the size of the frit. As mentioned many times previously, opacity and translucency vary with the mesh size of the frit used. Traditional pâte de verre works have a cloudy appearance. This is the result of having been made from very fine frit (>100 mesh) which traps a multitude of small air bubbles. The temperatures used in kiln casting are not high enough to allow these bubbles to escape.

These days, it seems that most artists are actually striving for that dime store transparency that comes from using a slightly larger frit size range. Using a number 10 to 14 mesh size frit will give this translucent look, while pieces made with larger chunks, "chunk de verre", or big pieces, "sheet de verre", will result in nearly clear castings. Be careful when adding large chunks of glass to a mold so, as not to scratch it, otherwise you will end up with flaws in the surface of your casting.

There is second possible effect achievable using a single multicolored mix of frit. Here different colors of frit are dry blended using a wooden dowel to give a spotted effect. You could probably refer to this process as "spotted dowel de verre" which is currently a rare and endangered art process. Use of mixes like this result in a finished casting with a look similar to marble.

Figure 106. Effect of using different sized frits.

Figure 107. Flowerpot in place to do a drip casting.

Figure 108. Effect from layering colored frits.

If you want to make a very fine-detailed, clear kiln casting, you can try the casting technique that Lucartha Kohler calls drip casting. Here you suspend a crucible with a hole in it above your mold. This can be done by setting up the crucible on some soft firebrick or on a kiln shelf with a hole in it. Fill the crucible with relatively large chunks of glass to minimize air bubbles in the melt. Then as the glass heats up and melts, it will flow down through the hole in the crucible into the mold waiting below.

Since the mold is at the same temperature or just slightly below that of the glass, the glass will more completely fill the mold. In addition, there will be no chill marks in the glass as seen in a typical casting process where the glass is rapidly chilled as it hits the mold. This method allows achieving very clear castings with finer detail than typically attained by kiln casting.

When casting with this technique, you usually have to put more glass in the crucible then is required to fill the mold because some of it will coat the crucible walls and not flow out. For this reason you will also often have overflow of the glass with this technique. This overflow can lead to a problem with cracking of the casting as the over flow clings to the mold, fire brick, etc. To alleviate this concern, it is suggested that you kilnwash the top of the mold and any materials such as firebricks that might be present. Also to protect your kiln you may want to add some whiting, sand, or powdered plaster on the kiln shelf around the mold to contain any overflow.

Kathleen Stevens uses drip casting techniques in producing her trees. Instead of using a crucible though, she uses flowerpots to melt her lead crystal in. These are a lot less expensive and with a little care to avoid thermal shock of the flowerpots, she can generally gets four to five firings from each one.

With some molds, especially bas-relief ones, it may be possible to easily fill different sections of the mold with different forms of loose frit. They may be of different mesh sizes to achieve opacity variations, or of different colors to add depth. Figure 106 shows how different mesh sizes of window glass frit were used to accent features on a zany cat head piece.

Figure 109. Fine color detail from precise frit placement.

For different colors, you could fill simply by pouring one color of frit after another into the mold to get a Neapolitan ice cream effect as in Figure 108. With this technique, you are never quite sure what you will get. As an example see the dark line midway up the question mark. It is due to an interaction between Bullseye yellow and turquoise. If you prefer to be more deliberate than this, you could place the frit more precisely as is done in sand painting. The Indian totem pole casting shown in Figure 109 was done this way. To see a real master of this technique, see the work of Ruth Brockmann.

When you cast a complete piece from different frit sizes or colors in a single firing, you will always get some blending of the glass as the frit shrinks in volume and moves

Advanced Casting Techniques

Table 9. Glass-modeling material density ratios.

	Soda-lime glass (2.5 gm/cc)	Lead glass (3.2 gm/cc)
Wax (0.9 gm/cc)	2.8	3.5
Clay (2.0 gm/cc)	1.25	1.6

about during consolidation as well as colorant interactions. Depending on how much mixing you are willing to accept, you might trade off mixing against how fully dense you want the final casting. In other words, you might try fusing at a slightly lower temperature to allow the grains to tack fuse together without fully consolidating. Here the final piece will end up with a texture like sugar or if the particles are large enough you may get "pâte de sushi." This is like the work of Karla Trinkley and Diana Hobson.

Determining how much frit to use

When getting ready to do your final casting, a lot of artists find themselves at a loss in determining exactly how much frit they will need to use to get a complete fill. This really doesn't need to be a mystery. There are actually quite a few ways to determine this.

One easy way to estimate how much frit will be needed to completely fill a mold was first suggested by a friend of Dan's, Scott Ingram. After cleaning out the mold and before drying it, you take a jar full of water and mark the water level. Then you quickly pour enough water into the mold to fill it to the desired casting level.

You have to fill the mold quickly otherwise water wicking away into the mold will cause you to overestimate the water needed to fill it. Presoaking the mold can minimize this error. The amount of water measures the volume of the mold cavity. Now by adding frit to the jar until the level comes back up to the original mark, you can measure an equivalent volume of frit. Then just dry out both the frit and the mold and you are ready to cook.

A second technique to determine how much frit to use is to measure the volume of the modeling material that you remove from the mold. If it is clay, dig it out and if it is wax, melt it out. Once you have it all, weigh it on a scale. Then multiply the weight of the modeling material, by the density ratio of your glass to your modeling material. Table 9 gives some approximate ratios that you can use. The product will give you the weight of frit that will need to add.

For example, if you remove a half-pound of clay from your mold and want to make a lead glass casting, then you multiply this by the lead glass to clay density as follows:

(lead glass weight)=(0.5 lb)*(1.6)= 0.8 lb

If you do not have a scale, you can measure the volume of the modeling material to determine the volume of frit to use. You can measure the volume by water displacement—measuring how high the water in a jar rises when you add the modeling material.

This method can be rather messy with water-based clays. To get around this we will often wrap the clay in plastic wrap. Also for models that you are going to burn out of a mold, you will need to do this before investing the model. Alternatively if the modeling material is flexible like clay, you can form it into a rectangle and calculate the volume. This is the length times the width times the height.

Once you have determined the model volume, you can measure an equal volume of frit again by water displacement. You have to use water displacement to measure the frit volume because its <u>pour volume</u> (the volume it has if you pour it from one vessel to another) or its <u>tap volume</u> (the volume it will have if you tap or vibrate the vessel it is in) are both much larger than the final casting volume. As we have mentioned, these volumes will be 30 to 50% greater than the final consolidated volume. This will also vary with the size distribution of your frit. The <u>water displacement volume</u> though will be exactly the same as the final consolidated volume.

If you don't want to deal with the wet frit, you can estimate an approximate volume of the frit required by multiplying the model volume by the frit volume multiplier of 1.47. This multiplier is based upon perfect packing of an ideal, single-sized, spherical frit. In reality, our frit is not single-sized or spherical and it is not perfectly packed. Instead all he particles are just randomly packed in the mold. Therefore the answer we get with this technique will slightly under estimate the amount of frit needed for casting.

Techniques for fritting glass

We have already mentioned one way of making frit in our chapter on the basic kiln casting process, this was the newspaper and hammer method. There are many other methods that are more productive than this. Let's look at some of them.

Thermal frit production

A good way to produce a roundish frit of fairly uniform size distribution is to melt some glass in a clay crucible and then slowly pour it into a large container of water as Dan is doing in Figure 111. The crucible that he is holding is a standard sized one available from most glass suppliers who sell fusing supplies. It is just the right size for this process.

Figure 111. Pouring a crucible of hot glass.

As the glass hits the water, it cools so rapidly that it can not contain all the stress within it and bursts apart into a fairly fine frit, on the order of 20 to 40 mesh. Because it can be tricky, we will discuss how to handle crucibles in our discussion on coloring glass.

Another method for producing frit with the heat from a kiln uses sheet glass. Here you start by piling up some of your compatible glass scrap on a kiln shelf. Then heat it to about 1450°F and allow it to tack fuse. Next don some heat protective attire. What kind is also discussed in the upcoming section on coloring glass.

Figure 110. Grinding frit with a mortar and pestle.

Then turn off your kiln (to avoid shocking experiences), reach into it with some long hot dog tongs, pull the fused pile out of the kiln, and dunk it into a large container of water. If the container is plastic, hold the glass off the bottom or it may melt through the bottom. The glass will immediately start breaking apart into frit.

This fritting method also works well, but produces a frit with wider size and shape distributions most of which is larger, in the 5 to 20 mesh range, than that produced from the crucible method. The frit size tends to increase with the thickness of the glass pile you started with because the glass does not chill as fast as with a thicker stack. If you want bigger chunks still, allow the hot or molten glass to air cool or thermal shock on a kiln shelf washed with separator.

Mechanical frit production techniques

Frit produced by any of these thermal methods can be broken down finer by using mechanical methods because it will have many fractures that have not completely propagated to the surface. You can wrap the glass in something like newspaper or canvas and <u>beat on it with a hammer</u> to break it apart as we discussed earlier. Wrapping it helps prevent producing many flying shards. You should wear safety goggles that completely shield the eyes just in case.

Each artist seems to have his or her favorite mechanical fritting method from industrial hammer mills down to a ball peen hammer in a coffee can. Dan likes one pound can of Folger's to 8 ounces of hammer. He can brew the coffee in the morning to wake up the red-eyed students and pound the can late at night to compete with the neighbor's rock and roll music.

Even finer frit can be made from thermally preconditioned glass by working it in a porcelain <u>mortar and pestle</u> although because of the tempering effect of the quick chill, it may be easier to further grind mechanically manufactured frit. To grind glass this way, start by filling your mortar about $1/3$ to $½$ full with glass frit. Then, while pressing down hard, rock the pestle from side to side about a half dozen times, stir and repeat. This tends to be a laborious process and hard on your hands. So you may want to wear some light work gloves to prevent getting blisters. With some perseverance, you can produce a couple pounds of fine, 100 mesh or better, frit in an hour.

Steinert manufactures a large stainless steel glass crusher, which is advertised as a heavy duty mortar and pestle for crushing glass. It is all stainless steel and has a pestle bowl 3-1/2" deep and 2-3/4" diameter in which you put the glass that you want to frit. The mortar fits into the bowl tightly and has a 7" handle with which you jam it down onto the glass. It

Advanced Casting Techniques

looks like a piston and a cylinder and sells for just under $80.

We have found it hard to get a good punch on the glass with this tool because the fit between the pestle and the bowl is too tight. The tight fit causes the air that is trying to escape as you press down on the pestle to slow it just like a shock absorber on a car. One possible way to fix the problem is to drill a couple of small vent holes near the lower edge of the cylinder to allow the air to escape. You can make your own version of one of these mashers by using pipe fittings that you can get from the hardware store although it may not be any cheaper. Frit made by mashing with iron implements will have a wide range of sizes and will also have metal impurities that need to be removed using a magnet.

Figure 112. Commercial glass crusher by Steinert.

If you're a garage-sale-diva or a flea-market-master, you can also pick up used <u>food blenders</u> and <u>garbage disposals</u> for fritting glass. Boyce Lundstrum recommends using a ½ horsepower hammer mill garbage disposal mounted in the lid of a 30-gallon metal drum. If you choose to go this route, you should definitely wear safety glasses, gloves, and a respirator while grinding your frit. Boyce connects a ½ horsepower shop vacuum to a hole in the side of the drum to pull dust out during operation. This should probably be a HEPA vacuum to prevent contaminating your breathing airspace with fine glass dust.

Boyce Lundstrum first starts the vacuum and then the disposal before dropping in small pieces of glass. About a half minute after he turns off the disposal, he removes the vacuum from the side of the drum and uses it to clean out rounded glass chunks that will no longer grind from the inside of the disposal.

If you fail to do this, you may have to occasionally clean obstructions out of the disposal. This is done by inserting a long flat head screwdriver into exit port and pushing against the underside of the hammer plate. Make sure that power is locked off before you stick your hand in it. We pull the plug. This mechanical crushing method produces a frit of which 60% is finer than 17 mesh. He finds that he has to stop every 10 to 15 minutes to prevent overheating the disposal.

For a real fine frit, use old blenders to grind it up. Do not use these blenders for food processing anymore unless you want to grind down your teeth. They also make really gritty margaritas.

Coloring clear glass with oxides

You can color clear glass using a technique similar to that used for crucible fritting. We usually make a bunch of frit of different colors at one time. We start by preheating the crucible in the kiln. Then when everything is hot, we add the dry mixed frit and colorants to the crucible. This hot loading method seems to allow faster chemical burn off if it is going to occur.

Since you will be melting, pouring, and recharging the crucibles a number of times in a fritting session, some glass will inevitable run down the outside of the crucible. Some hot glass may also be spilled from the crucible inside or outside of the kiln if not handled carefully. Because hot glass is very corrosive, it can eat into kiln shelves, even when kiln washed, and will drill holes into soft brick or fiber blanket. For this reason, we like to lay about three quarters of an inch of silica sand directly onto the bottom of the kiln as shown in Figure 114. We fire the kiln without a shelf so that the crucibles stand directly on the kiln floor.

This may sound appalling to some kiln owners, but it is a whole lot easier cleaning some fused sand out of the kiln than replacing kiln bricking. Especially if the bricks are cemented in place—as most are these days. The sand will not melt since, as we discussed earlier, no glass kiln on the market gets hot enough to melt fused silica unless the sand is highly fluxed. But if a spill happens, the sand will diffuse it and prevent the molten glass from reaching the brick floor of the kiln. The sand also adds some stability in standing the crucibles up, since they are slightly conical in shape with fairly small bottoms.

When pouring crucibles from a kiln, a front loading kiln is the preferred configuration over a top loading one because it allows easier gripping of the crucible. So for someone who is going to do a lot of crucible work, such a kiln might make a desirable addition to your studio.

All activity, except heating of the glass takes place outside of the kiln. The crucible is best handled with some scissors-type, fireplace log tongs. Use these to pull the crucible out of the kiln, to pour it, or to refill it with a new colored frit mix. Using the log tongs may take a little practice but it is not difficult. If you have any doubts, try a couple of practice runs with a cold kiln.

For better control, you may want to consider wearing only one of the soon to be suggested gloves. Being right handed, we find that using a gloved left hand and an ungloved right one works best for us as shown back in Figure 111. Then by approaching the kiln from the left side, the gloved hand can help protect the bare one. The left hand acts mainly as a fulcrum point, a weight support, and a point to add a little extra squeeze on the tongs. It is the right hand, which really controls how much squeeze that you put on the crucible with the tongs.

Because the crucibles are slightly conical, it is best to approach them from the bottom with the jaws of the tongs open as you try to approach them. Then slide the tongs slightly upward and grab at about the middle of the crucible.

The intense heat coming out of the open kiln may be a new experience for you and it can cause some people who aren't used to it to speed up their actions. Try not to let this happen as it can lead to making a mistake and possibly to a serious accident. Take your time, breathe through your nose and be sure to keep a commanding grip on the crucible. Use your body and gloved left hand to shield your right hand from the heat of the kiln.

Since melting and handling crucibles requires the highest temperature exposures you are likely to receive during casting, we would like to suggest some basic safety practices to be followed.

1. First and foremost, anytime that you go into an electric kiln, other than to just see the show, turn off the kiln. The heating elements are hot in two ways: thermally and electrically. If you should accidentally contact an electrical element with a metal tool, you can get welded to the concrete and support a hairstyle by General Electric—

Figure 113. Aluminized coat for those really hot kilns.

assuming that you live through it. We don't understand why every kiln manufacturer does not have a deadman's switch on the door that turns off the power whenever the kiln lid or door is opened or enclose their elements in quartz tubing.

2. When doing crucible work, practice the buddy system. You will need that person to turn off the power and to open the kiln.

3. Never wear any polyester clothing or you may find yourself getting shrink-wrapped by it. Cotton works much better. Since you are getting radiantly heated when going in to pick up the crucible, it is best to wear long-sleeved cotton clothes. A good denim jacket is best, an aluminized fireman's coat like Dan is wearing in Figure 113 would be the best you could do.

4. When you go into the kiln, you are exposing yourself to intense UV and IR radiation that can severely damage your eyes with repeated exposure. You need to wear some sort of protective glasses. We will talk more about this later in our safety chapter but suggest that a welder's green of at least a #3 shade is appropriate. A #4 shade would be better but you don't want them so dark that you cannot see what you are doing once out of the kiln.

5. High temperature gloves are a must.

To many people, asbestos immediately comes to mind in one way or another when talking about high temperature gloves. Asbestos is fibrous form of magnesium silicate. It is not toxic but the short fiber form, can be inhaled and lodge in the lungs. There it can act as an irritant—not a poison but a carcinogen. Asbestos gloves offer great protection from heat but the long fiber can break down with repeated heat exposure to form the bad short fibers. For this reason, it is best to avoid asbestos gloves, even if you can find them.

There are other heat resisting gloves on the market based on Kevlar. Kevlar is basically a low

Advanced Casting Techniques 115

Figure 114. Hot crucibles set up in a kiln.

expansion version of fiberglass that can withstand heat to about 1000°F, for short periods of time. These gloves can be found at ceramics and welding suppliers under the name of Kevlar but for some reason are found at fusing suppliers under the name of "Zetex." What's in a name anyway. The important thing is that they should be double lined. A wool inner lining is the best.

For better durability and higher heat resistance, PBI/Kevlar gloves are best, although of course they are more expensive. Pure PBI (polybenzimidazole) gloves are available but their price is out of sight and you don't really need something so high-tech.

Terry cloth gloves are all right for low temperatures (under 500°F) but they will flash into flames at the temperatures that we are talking about for crucible handling. Also avoid leather gloves, because once they get hot, they stay hot, and that can be rather unpleasant.

Now that you're all dressed up and have had a couple of practice runs, you're ready for the big event. Use two five-gallon buckets with one filled about three-quarters of the way with water. Plastic buckets are okay for us but some people have had problems with them melting or scarring and for that reason have switched to large stainless steel mixing bowls in which to frit their glass.

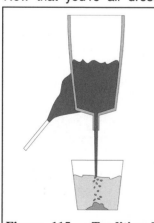

Figure 115. Traditional setup for fritting glazes.

These will not melt nor will they allow the glass to get embedded in the bottom.

So pick up the molten glass filled crucible out of the kiln with the tongs (don't forget to turn off the power first) and empty the crucible into the water. Go slowly at first because glass is viscous and tends to want to spill out as a big lump. A big lump could burn a hole in the bucket, which can then flood the studio floor (not a good thing when working with electrical kilns) besides ruining a good bucket.

After you are done pouring, fill the crucible with the next batch of glass and get it back into the kiln. If that was your only or last batch, still put the crucible back into the kiln and let it cool gradually or it may thermal shock and crack on you.

After completing the pour and replacing the crucible into the kiln, you can pour off the water from the one bucket into the other bucket and dump the frit at the bottom onto some newspaper to dry. Repeat the process after each pour to avoid mixing colors, wiping out any fine frit residue with a paper towel. If you don't like the colors that you are getting at least you will have a couple of gallons of warm water.

You can also color your glass and frit it as the old studio potters' used to make their fritted glazes. Here you mix up your raw materials (glass cullet and colorants) in a funnel shaped crucible with a small hole at the bottom. The crucible is suspended over a bucket of water. A flame directed on the outside of the crucible as illustrated in Figure 115 heats up the mix to melt. Once molten, the mix flows down through the hole in the bottom of the crucible and drips into the bucket of water. The sudden change in temperature instantly freezes the drops of glass and shatters them into a fine frit. Be sure to provide plenty of ventilation if you try this because of the high toxicity of many colorants.

Commercial Glass Frits

Frit is available commercially from a number of sources. By far, the soda-lime glass frit used most often by studio glass artists with America is that from Bullseye Glass Company. It is available in the wide color palette and compatibility quality control for which Bullseye's fusing product line is known. It is comes packaged in either 1½-ounce tubes or five-pound jars. The 5-lb. jars were seen pictured back in Figure 28. The available frit sizes they carry are listed in Table 10. They also have a clear non-lead

Table 10. Frit sizes available from Bullseye Glass.

Frit Designator	Size Description	Mesh Size	Particle Size (in)	Particle Size (mm)
01	Fine	14-70	0.008-0.047	0.2-1.2
02	Medium	10-14	0.047-0.106	1.2-2.7
03	Coarse	5-10	0.106-0.205	2.7-5.2
08	Powder	> 70	< 0.008	<0.2

Table 11. Frit sizes available from Uroborus Glass.

Frit Designator	Size Description	Mesh Size	Particle Size (in)	Particle Size (mm)
F1	Powder	>60	0.005-0.01	0.13-0.25
F2	Fine	23-60	0.01-0.03	0.25-0.75
F3	Medium	8-23	0.03-0.1	0.75-2.5
F4	Coarse	2-8	0.1-0.3	2.5-7.6
F5	Mosaic	-	0.3-0.6	7.6-15

crystal with negligible iron content which is available in either billet or chunk form.

Uroborus is now also manufacturing frits and powders that are compatible with Bullseye. The sizes of frit that they have available are listed in Table 11. They are also manufacturing COE 96 frits that are compatible with Spectrum Glass' tested fusing line and may even be fritting some of Spectrum's glass.

The lead glass frits are mainly available from ceramic supply houses. For a possible supplier, you could try Standard Ceramic Supply. They still carried them when we started writing this book. They are getting harder and harder to find though.

Traditional glass pastes

In molds with a lot of slope, it is not always possible to fill different sections of the mold with different colors of loose frit because it just slides down the sides. In this situation, you must make use of the full potential of the pâte de verre process, i.e. make a paste of your glass. Here one uses the technique as it was developed and used by the French masters. With this technique, you can achieve subtle shading of colors. To be able to apply your glass paste with control, you need a mold with large enough openings to allow complete access to all surfaces.

Making glass pastes

You make a glass paste by mixing your frit with some water or water-based binding agents like those used in glass painting, pottery glazes, or enameling. The paste can then be brushed or packed it into place in your mold with a palette knife. When mixing up your paste you want to mix it to a smooth consistency like that of a thick paint. It should be easily picked up with a brush but at the same time it should be able to just as easily drip off of it.

Water alone can be used to bind the paste and burns out clear (at least in most cities) but lacks the pastiness necessary to hold the glass frit in place on near-vertical surfaces. Oils as binders are not generally recommended because they can carbonize and there is usually no place for the carbon to escape as the oil burns.

Of course there are exceptions to every rule. Donna Milliron has told us that she has had good success using squeegee oil which hardens almost like a glue and has not had any problems with either carburizing or color reduction. If you are going to use oil-based binders it is advisable to use them sparingly, to add a soak at 1000°F to burn them out during the firing and to have a fairly permeable or porous mold to prevent trapping organic vapors.

Traditionally one part of gum arabic (from the acacia tree) dissolved in one part of alcohol and then mixed with twelve to twenty parts of warm water was used as a paste binder. Be careful with this though because it can also carbonize and "fry," that is cause bubbling or crackling in the glass, if the solution is mixed too strong. It's a pretty strong binder.

Other traditional gum binders include: gum tragacanth, dextrine, vee gum and CMC (sodium carboxymethyl-cellulose). They are all mixed similarly to gum arabic and can be found at many ceramics supply stores. If you can not find anything

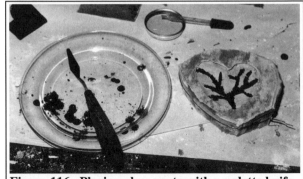

Figure 116. Placing glass paste with a palette knife.

else you can always thin down some Elmers glue with water and use that to hold your frit in place. It works just fine.

A good binder found around most households is a diluted solution of Knox gelatin. It doesn't take much gelatin to get the glass to stick and you can even drink the leftovers to get those chrome-molly fingernails we talked about earlier. Another simple water-based binding agent is sodium silicate dissolved in water. Any of these mixtures will harden when air-dried so you should only mix as much as you need.

Thompson Enamels makes a good binding agent called Klyr-Fire. This is a methyl-ethyl-cellulose based binder that has given us good results. It burns out clean, is cheap and you don't need a letter from the governor to buy it directly from the manufacturer or its distributors. No mixing is required and it can be used straight from the bottle. Brushes used with it can be cleaned with water afterwards. The paste does not harden as much as some of the other mixes though and may move around some on you.

Glass pastes painted into your mold can be used to achieve very subtle shadings of color through the use of a palette of colored frit or incorporation of metal oxides or enamels. The addition of 10 to 20% by volume of 80 to 120 mesh enamels is sufficient to create sufficient color density when used with a clear frit.

Enamels work best with the low-temperature, high-lead (20 to 25%) glasses because their flex points are similar. Otherwise you may have to adjust the coefficient of thermal expansion of the enamel to fit the base glass frit that you are using. This can be done by mixing it with the desired base glass, melting the mixture in a crucible, and refritting it by pouring the melt into water. Make sure that you mix the component well before starting and allow the melt to sit at high temperature for a little while to allow good mixing of the components.

Color blending glass pastes

As you work in kiln casting you will often find that you want custom colors. The glass core of your casting will usually look better if made of lighter tints of colors since the mass of the glass will intensify the effect. Bullseye Glass, recognizing this fact, has added a small line of transparent caster's frit with lower color intensity.

Don't be afraid to use those wimpy colors for the core of your work since too strong a color will render the whole piece opaque, unless of course that is the effect that you are trying to achieve. One exception with using mainly more transparent glasses is when you are casting a hollow vessel like a cup, bowl or vase. Here the thin walls and the light entering through the center of the vessel allow use of richer colors. This is also true of objects like lampshades, which are going to be illuminated.

Some artists find that they want to dilute their commercial transparent colors to get the color saturation that they want in their work. Some do this by melting a little transparent colored glass in a crucible to which clear glass has been added and then refritting the melt as has been discussed earlier.

Others, like Jim, rely on mixing transparent powders with some medium ground clear frit. Using a binder like Kylnfire will allow the powder to evenly coat the clear frit. Mixed properly this can be used achieve a good uniform color when fired. This is sort of like mixing color paint at the hardware store. You have to be sure to mix the powder pigments and the clear glass well to prevent color variations.

When mixing colored powders or enamels with clear glass frit, you can sometimes have a problem getting a uniform mixture of color. If the clear frit is much larger than sugar crystals, you can get separation of the two. We find that it works best if you wet the clear glass with a binder first before mixing in colored powder. Add the powder slowly so that it does not lump up. This will allow the powder to evenly coat the frit. By watching how the color of the mixed powders is changing, you can also get a feel for how the depth of the color in the finished product will turn out.

When mixing the powder with the frit, realize that the amount of powder that you add will affect both the color and the translucency of the casting. So if you are trying to get richer color without decreasing the translucency, add a little darker color of powder rather than adding more of a lighter color. You can also decrease the translucency of a casting without increasing the depth of color by adding more of a lighter colored powder. You can even change the translucency of a casting without affecting the color at all by adding clear powder into the paste.

Since we use mainly Bullseye frit, we work with Bullseye and Uroborus powders for our pigment glass. Other artists use Spectrum Glass and Spectrum-compatible Uroborus frits. You could also use Kugler and Zimmerman powders. These are matched well with Schott glass. Because you usually use so little powder relative to the base glass and it gets so well mixed into the glass frit, you do not always have to be as careful about compatibility of the powders. But we still recommend not mixing different types of glasses to avoid compatibility problems.

Since this is such a powerful technique, you will probably want to work with it to develop your own custom colors. We would suggest primarily using a single colored powder when mixing up a paste this way. This is because colored powders do not necessarily work like paints. The metal oxides can interact to give some really ugly colors. But this does not mean you might not get pleasant surprises. It's just that things are more fool proof if you stick to a single powder.

Figure 118. Mold for casting color wedge samples.

You will want to get a feeling for how the depth of color will change with both the weight ratio of colored powder to clear frit and the thickness of the piece. To do this, you should make up a color sample set to see how these factors interact. The best kind of sample for this is to cast wedges of glass because this will let you see how the depth of color changes as the thickness of the glass increases.

Jim has made up the set of strong investment molds with good draft as seen in Figure 118 that he uses to cast his color sample edges. By using a little kiln wash on the inside of the molds, they can be used for multiple casting runs. The glass wedges he makes are about an inch and a half wide by three inches long. They taper from an eighth of an inch thick to three-quarters of an inch thick. The molds have the same shape with an extra half-inch approximately rectangular section added above the wedge. It is "approximately" rectangular because of the needed draft. Jim has it worked out exactly how much glass is needed to fill the molds and thus cast repeatable wedges.

Cast some up yourself to explore the effect of the powder-to-clear frit ratio. At the same time, you might want to explore the effect of the clear frit of different mesh sizes. Look at how the color mix effect and the translucency varies.

Using glass pastes

Once mixed, the glass paste is painted on with a brush in thin layers into the low relief areas of the mold. A small flexible palette knife also makes a good tool for paste application. Remember that because the mold is a three-dimensional negative of the finished work, these low relief areas will end up being raised areas in the final piece.

Fill in these carved out or detail areas first using whatever color your design dictates, one color at a time. Let each paste color dry and harden before you move on to the next. This will help minimize color mixing of the pastes. Then continue applying paste until you have the entire surface of the inside of your mold covered to at least $1/16^{th}$ of an inch thick.

Sometimes it's a little tricky to get the paste into the neither regions of a mold and it may require considerable dexterity. A mini-mag flashlight can be helpful in exploring those really deep molds. It also helps to keep the mold wet by wrapping it with a damp towel while applying your glass paste as in Figure 117. This prevents the water-based binder from drying too quickly. This technique will increase the working pot life of your paste, give it more strength and decrease your chances of forming any air bubbles.

Figure 117. Wrapping mold during frit application.

Advanced Casting Techniques   119

Figure 119. Hollow vessel mold packed and ready.

After application of the first layer of paste, let it dry to avoid disturbing it as you build up the thickness of the rest of the piece. You can let it air dry naturally or use a hairdryer if you are in a hurry. Then build up the thickness another 1/16th of an inch to about an eighth of an inch by applying color as desired to build color density. Many times, the second layer of paste is used to form a lighter background or semi-transparent color to prevent the piece from becoming too dark or opaque.

Try to gently pack the paste down to decrease movement during the fusing process. To help minimize color any motion and mixing, it is recommended that the first two layers be fired to consolidation by going to the light tack fuse process temperature for three to four hours depending on the thickness of the mold. By then the glass will have shrunk to about 70% of its original volume. This is easily done if the mold does not have too vertical of sides.

For highly sloping objects or hollow vessels, you will need to pack the inner surface with ceramic filter paper backed with ludo as shown in Figure 119 or insert an inner mold piece for a two-part mold. Put a weight on the inner piece to drive it in as the frit volume shrinks. This will minimize sliding of the paste in the mold as it consolidates.

In the past, artists such as G. Argy-Rousseau, would pack the center of the vessel with asbestos fiber. That technique is not practiced now-a-days for obvious reasons. For similar reasons, many people are also becoming concerned about the health risks of using ceramic fiber paper.

After firing the first layers, additional layers are added and fired until you get the final thickness in which you are interested. In the last firing, you can continue up to full fuse temperature to get a full density casting.

Again there are really no hard and fast rules in kiln casting. By careful color placement and buildup, many glass artists are pleased with the results that they achieve from a single firing. To avoid problems of color movement, they will construct molds without steep sides or work mainly with homogeneous frits.

In choosing the placement of your colors, it is often advisable to use your darkest and richest, or most opaque colors for the relief detailed sections as they will show up the best. Be careful in their application, because in thicker layers the light can be obscured to such an extent that the piece can end up looking like one big completely opaque dark mess.

Inclusions

As discussed previously, you can combine homogeneous frit cast pieces together by reinvesting them in a mold, adding filler frit, and recasting the pieces to join them into multicolored pieces. Of course, you are not limited to single-colored components with this technique. You are also not limited to just joining them.

Why not use one casting as an inclusion inside of a larger work? You could make fish swimming in a fish bowl. When you use this technique, you want to be sure that your outer glass is translucent enough to be able to see the included object in the casting. For this you want the outer frit to be a very transparent color and a larger mesh size.

You can also incorporate fused elements or trailings poured from crucibles. David Ruth likes to use a number of glass elements like this in his large sculptural pieces. He feels that they give his work more depth. They act as focal points to draw the viewer into the piece. You do not need to restrict yourself to cast or fused glass inclusions.

Many of you may have some skill with lampworking. Why not use that skill to make lampworked objects to use as inclusions. This is what Donna Milliron does with her pâte de verre flower beads. This concept can lead to many interesting kiln casting possibilities. How about making lampworked insects that you then cast as inclusions in an amber colored glass? If you want to be even more playful you could make small, encased dinosaurs and sell them as dinosaur theme park starter kits.

You also do not need to limit yourself to glass inclusions. Thin metal foils, leaf and screens will also work well. Try to make sure that they are completely encased and keep them small if you want to avoid cracking.

Introducing shaped voids

Another kind of inclusion that you may want to consider adding into your work is negative inclusions. This is an area that ends being a void in the final piece. These are formed by investment inclusions that are partially enclosed by your casting that you remove post casting to make a hollow space. One form of this that we have already discussed is the drag in a hollow vessel mold.

This technique can be used for more than just making hollow vessels. They can be used to put decorative images into a casting. These images will be visible from the exterior of the casting and will look like objects inside of them—especially if you decide to paint the cavity.

Some of you may decide that you like the look of the investment inside the work and not want to bother to remove it. We would advise you to remove it because it can act as a source of stress in the glass that could cause it eventually to crack. This can especially be a problem if you get the investment wet. This will cause the investment to swell and exert more force on the glass. If you are thinking about retaining the investment try to seal its exterior surface with a coating of lacquer.

If you instead decide to remove the investment as we recommend, you remove it just as you remove a drag from a hollow vessel mold by digging it out. Dig it out little by little using sharp pointed tools like an awl or a chisel. Do not drive the tool deep down into the center of the investment to break it up as this will force all the investment out against the glass and could cause it to crack.

For the same reason, do not try to soak the complete image out of the glass. As we discussed above, the investment will swell as it absorbs the water and exert force on the glass. Instead, dig the investment out piece by piece. Work around the edges of the investment but try not to hit the glass with the tool as this can chip or scratch it. You can wash the last of it out with a toothbrush and water.

You can make your investment images by carving them out of a rough blob of investment cast as part of the mold or separately model it as was described for making plaster models in the modeling chapter. If you decide to do this, you will want to work the investment wet and ensure that it does not include any chopped fibers in its formulation.

Figure 120. Casting a investment inclusion.

The other alternative is to cast your investment images using master molds as shown in Figure 120. You can use the same kind of flexible master molds that we used for making wax replicas. Do not try to achieve the same kind of delicacy with investment that you can get with wax. Investment images have to have sufficient thickness to them to be able to demold them without cracking them. This is made easier if you applied a little release agent to the inside of the master mold.

To incorporate a stand-alone investment image, attach it to a wet mold using a little thinned investment as a glue.

Kiln Procedures

The topic of this chapter is kiln operations. The two main kiln operations in kiln casting are mold burnout and casting firing. Either of these operations can be broken down into three phases: heating, process, and cooling. We will focus on each of these in great detail in this chapter. We will also provide you some general rules of thumb for determining the parameters to use in each phase. After that we will finish the chapter with a discussion of different pieces of equipment for measuring the temperature of and for controlling your kiln.

Kiln operations

If you have been a glass fuser, one of the things that you will find interesting in our discussion of kiln operations is how in fusing and slumping we go to great lengths worrying about thermal shocking the glass and don't worry about the molds. In kiln casting, exactly the opposite is the case. The molds are the vulnerable components and not the glass. Gypsum plaster-based investments weaken considerably as they are taken to high temperatures and may crack if abused. That's why we went to the trouble to include refractory materials and property modifiers in the investment formulations listed previously.

So let's look at the first of the kiln operations we perform in kiln casting—mold burnout. Burnout is done to remove any organic remains from your model that might be left in the mold be it wax, wood, plastic, or some other material. After that we will discuss the other kiln operation—casting firing. This transforms the frit encased in the investment mold into the final piece of kiln cast art.

Curing and burning out of molds

After melting or steaming out any wax, air drying, and reinforcing the mold, the next step in the kiln casting process is often to heat cure it or burn it out. If you have been successful in completely removing all modeling material manually or by steaming it out as in case of wax, then these important steps can be included as part of firing the casting.

What we are concerned with in curing a mold is removing all the water and organic materials without cracking it. As explained earlier, any water remaining in a mold may be transformed into steam by the heat of the kiln. This transformation involves a tremendous change in volume that has to be resisted by the strength of the investment material.

If not done carefully, this transformation can result in cracking of the mold in a similar manner as to how volume-expanding epoxies are used to break up granite boulders as an alternative to dynamite when building roads in the mountains. So to prevent cracking the mold, we want to remove this water slowly.

Water exists in the mold in two forms: physically adsorbed water and chemically bound water.

Physically adsorbed water is that water which remains on the surface of the investment crystals after the casting process or has been attracted from the air. This water is loosely attached and can be removed slowly by air drying or more quickly by heating the mold to around the boiling point of water. At this temperature, any loosely bound water is rapidly transformed to vapor. If you have thoroughly air-dried your mold, you may have already accomplished this task and this portion of the cure cycle can be minimized.

Chemically bound water, on the other hand, is that water which actually becomes part of the chemical makeup of the investment. As in the manufacture of gypsum plasters and cements, the investment mold has to be heated up to at least 350°F to drive this water off. Therefore when curing a mold, dwells at both 225°F and 350°F are included during the heating phase to carefully remove both of these forms of water.

The other part of this operational discussion, burning out all organic remains in the mold, is done so that they do not coat the glass that you are casting with carbon. To completely decompose most organic materials, you have to heat them to temperatures of about 1000°F and sometimes as high as 1200°F. Make sure that the studio and kiln are well vented when doing this because there could be some smoke.

If you were real diligent about steaming out your wax and don't have restricted access cavities to trap wax, then you probably will not have enough wax left in the matrix of the mold to worry about carbonizing. But to minimize any chance of problems, you may want to include a dwell at a temperature between 1000 and 1200°F to burn off all carbon before proceeding on to casting temperatures. This will leave a clean mold with wax but other dense organic materials like woods or vegetables may leave ash behind after burnout. Any such ash will have to be cleaned out of the investment mold before filling it with glass frit.

Firing a casting

When you have cleaned your mold and filled it with glass frit, you are ready to fire it to consolidate your glass. Figure 121 shows a kiln setup with molds ready to be fired. In this operation, if you have not previously heat cured or burnt out your mold, you will have to be concerned about any moisture that might still be present in the investment as was discussed above. In addition, you will have to be concerned about exposing the glass to the proper time-temperature conditions to achieve the final desired product.

As any good glass fuser knows, your final results will be a function of both the processing temperature as well as the length of time you spend there. One can be traded off for the other. Here the relationship will be a little different from what you may be used to for fusing because of the insulating nature of the investment mold and the greater amount of flow that is necessary. More time has to be allowed for the glass to change temperature after a change in kiln temperature.

Figure 121. Kiln full of filled molds ready for firing.

The process temperature that you should use for kiln casting is a function both of the type of glass that you are using and the degree of fusion desired. Higher temperatures are required for harder glasses and for greater casting density. Remember that darker colored glass absorbs heat better and appear softer.

Three phases of any kiln operation

Any kiln operation is basically divided into three phases: heating, process, and cooling. This is seen in the typical pâte de verre firing schedule illustrated in Figure 122 that we repeat here from our introductory chapter.

The heating phase for the most part is pretty self explanatory and our biggest concern in that phase is preventing thermal shock of the mold. The process phase is that portion of the firing in which the process, in this case kiln casting, is accomplished. Here the concern is getting the proper heat work into the glass to fully form and fuse the casting. Lastly is the cooling phase. This is considered by many kiln casting artists to be the most critical phase because it is here where the glass must be carefully returned to room temperature in a minimally stressed condition to avoid possible cracking of the final work. This seems to be one of the least understood and easiest things to mess up.

Let's look at each of these phases in greater detail.

Heating phase

In the heating phase of a glass fusing firing, you are concerned about not heating your fusing piece so fast that you thermal shock the glass in your project. Thermal shock is breakage that is the result of stress building up from variations in temperature throughout the piece of glass and the corresponding relative expansion it undergoes. Here expansion of the hotter edges of the glass can pull at the cooler inner section such that you literally pull the glass apart.

Breakage of the glass is not usually a concern in the heating phase with kiln casting because the glass is already in the form of frit rather than large pieces. In addition, the thermal insulation of the mold slows the rate of temperature rise of the glass and smoothes out any temperature gradients.

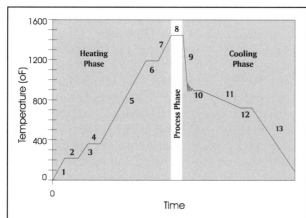

Figure 122. Three phases of a kiln operation. The rates and soak times of the segments will vary with the size of the project as explained in this chapter.

In kiln casting, the only time that this tends to be a problem is in refiring of a large bas relief casting still in the mold in an attempt to fill it up some more or when you are fire polishing a casting. In either case, the thinner areas of the glass casting can heat up sufficiently faster than the thicker area or the areas buried deeper in the mold to cause them to crack. But if the piece is still in the mold and you are just heating it up to fusing again anyway, it may not matter.

Instead for kiln casting, we need to be concerned about the mold. We want to avoid abusing it. So for that reason, we will assume that it still contains considerable moisture and maybe some organic materials in it. This discussion can also be applied to a filled mold that did not require a separate burnout cycle.

With this in mind, let's discuss how you go about heating these puppies up to cure and burn them out. Start by putting them into the kiln with the reservoir opening up. Leave at least an inch between molds to allow good air circulation between them. Then close the kiln but leave the top propped open about an inch with a piece of kiln furniture.

In the first segment of a pâte de verre firing, **segment 1,** as illustrated in Figure 122, we want to heat the kiln slowly to about 225°F.

Now comes that age-old question; how slow is slowly? This depends upon a lot of things.
- First, how wet is your mold? Did you do a good job air drying it?
- Second, how big is it? Are you modeling a small bug or King Kong?
- Third, how much time have you already invested in it and are you willing to risk having to start over?
- Fourth how strong is it? Did you use a reinforced investment?
- Lastly, are you a gambler or are you afraid to even test your luck at the lottery?

If you answer all these questions and mail them in on a post card, we'll be glad to misinform you.

Slowly is something that you will have to decide. We would suggest that you keep your heating rate in the neighborhood of 50 to 300°F per hour per inch thickness of the mold. Most of the molds that we construct are only about an inch thick at the thickest point, so we set the controller to reach 225°F somewhere between 30 minutes and three hours. If your mold were twice that thickness, you would want to go slower by half that rate or take twice as long, about an hour to six hours, to get to 225°F.

Once you reach 225°F, you will next dwell as in **segment 2** at this temperature long enough to allow the mold and the glass to come to temperature, all the water to evaporate, and all the vapor to be transported out through the porous structure of the outer surface of the mold.

You're not going to ask how long is long enough again are you? Go back and look at your answers to those previous questions. The same factors apply for this segment as the last with the addition of how porous your mold is. We suggest that long enough will usually lie within 1 to 2 hours per inch thickness of mold. Again, if you have thoroughly air dried your mold, this portion of the cure cycle can be shortened. If you have a non-porous container around your mold, such as a flask, this dwell and the next should be increased.

Now that you have removed all the physically attached water, your mold is through the most likely cracking region. But we suggest that you continue heating slowly in next segment, **segment 3**, up to 350°F where we stop to remove the chemically attached water. Let's keep the same definition of slowly as in segment 1, so it will take about the same amount of time to reach this temperature as it did to reach 225°F.

In **segment 4** dwell at 350°F for a little shorter amount of time than that of segment 2, say on the order of ½ to 1½ hours per inch thickness of the mold. After this dwell, you should have removed all the chemically bound water and you should be able to heat faster from now on.

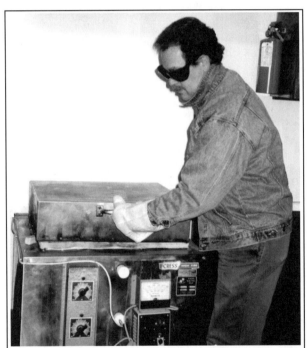

Figure 123. Proper protective clothing for kiln entry.

In **segment 5,** you raise the temperature in the kiln to burn out all organic materials. You can speed up this segment and the next heating ramp. Doubling the rate to about 100 to 600°F per hour range per inch of mold thickness should work okay. Mold thermal shock, although not likely at this point, is possible so don't push the pedal to the metal too hard.

Any organic materials will start to burn at about 500°F and produce smoke. This means that it is time for everyone to go to lunch. At one workshop Dan gave, the kiln was in the employee lunchroom. His class went out for lunch, but the employees ate in. As you can guess, he was not the most popular out-of-town expert that day. Obviously, this should not be a common problem because we are sure that most of you follow good work practices and do not eat, drink, or smoke in your studio anymore often than you burn out wax in the employee lunch room.

Once the kiln gets to about 1000°F, most of the smoke will be gone but burn out may not be quite complete. If you open the kiln for a peek, the molds may "candle" or burn with a small flame. This means that there is still some wax in there. You may also notice that the mold has a black or gray carbon build up especially around the reservoir opening. This carbon is what remains from any organic materials that were in the mold. The molds have to be cooked until the residue completely disappears leaving a totally white mold. Otherwise any residual carbon can fuse onto the surface of your glass and will be very hard to remove.

In **segment 6**, you should soak at this temperature or at most about 1200°F. Sometimes the soak at the top of this firing can last as long as several hours before you burn out all the carbon. This can be due to wax having wicked into the mold especially if it were melted out rather than steamed out. Continue to dwell at this temperature until the mold appears clean.

Once the mold appears clean and white you can proceed in **segment 7** to continue heating up to the process temperature. The suggested heating rate here is the same as in segment 5.

If the mold was empty when you started, you have a choice to make once burn out of the mold is complete and before proceeding on to segment 7. This assumes of course that there is no ash that needs to be cleaned out in which case you are forced to cool the mold back down to room temperature and remove the ash.

The choice you have is between filling the mold with frit or whatever size of glass pieces you prefer on the spot and turning this into a casting firing, or cooling the mold back down and filling it at your leisure.

The first choice would save having to do another firing, but in return you forfeit the chance for any color placement. To fill the mold when hot, something like a lead ladle wired to a length of half-inch re-bar for stained glass is suggested. Most of you may have some old re-bar laying around since you were probably into stained glass at one time. Lead ladles may be a little harder to find since they are being unofficially considered as undesirable lead abuse paraphernalia. Try a plumbing supply shop or hardware store. Otherwise consider using a large serving spoon or non-aluminum metal scoop from the kitchen.

You may need these ladle or spoon rigs later anyway for topping off the glass in your reservoir at the top end of the casting firing if your reservoir is not big enough. Try to get in and out of the kiln fairly quickly or you risk thermal shocking the mold. It is relatively weak at this point and a cold blast of air is the last thing that it needs.

When opening the kiln at these temperatures, you have to be careful because of the extreme heat and all the electromagnetic radiation coming out of it. Besides making sure that the power to the kiln is turned off, make sure that you are not wearing any polyester clothing. If it gets too hot, this material melts and can do a shrink-wrap routine on you, which we guarantee will not be pleasant. It does not cool off very fast afterwards either and will continue to burn you.

Wear cotton clothing instead because it will start smoking long enough before anything worse happens that even we can tell that we have been "too long at the fair." Your protective clothing should be long sleeved of course. An old blue jean jacket is ideal. Also use high temperature gloves that are from the Kevlar family. Avoid asbestos. Figure 125 shows Jim dressed out for kiln entry.

Instead of filling the mold at this point, most artists prefer to cool them back to room temperature so that they can clean out any ash and carefully compose the color placement of their glass pastes, frit, or chunks into the mold. This is where a lot of the real art in pâte de verre takes place. Cooling the mold should also be done with some care, because now that the chemically attached water has been driven off, the mold will now be rather fragile and must be treated with respect. It is vulnerable to thermal shock and will have to be cooled slowly.

Oh! Oh! There's that ambiguous word again. Cooling in the range of 100 to 600°F per hour per inch of mold thickness should be okay. You can go on the fast end with small molds but take your time for those big boys. If you hear any pinging sounds while peeking at the molds in the vented kiln, this is probably indicative of micro thermal shocks in the molds that may lead to hairline fractures of the mold. These can further develop into cracks that may leave noticeable flashing lines in your finished work if they get very large.

All this was fine for a wet mold, but what about a reinforced one that has already had a heat cure cycle or was well dried. If your mold has sat around for a long time since heat cure, you might want to consider that it may have picked up physical water and give it a similar heating cycle to that for a wet mold. If not, you can do away with all this and heat right up to the process temperature without any stops along the way. This rate should be in the range of 100 to 600°F/hour/inch of investment thickness.

Also if you were able to steam most of the wax out, you may want to have only one kiln cycle and combine burnout and firing. Here you could heat the wet mold in the normal way to a burnout hold (segment 6) and spend an hour or so before going on to heating in segment 7 to the ultimate process step (segment 8).

You may want to make some test runs to see if you can live with whatever wax you may have left in the molds after melt or steam out to see if it will affect your work, It is really a lot easier if you only have to fire your molds once and can skip having a separate burn out run. Many artists find that this works for them.

Process phase

In **segment 8** or the process phase, we're still worried about the health and well being of the mold since you still can't do much to harm the glass. The mold is getting weaker and weaker the higher we take it in temperature. For this reason, pâte de verre has been traditionally done with highly fluxed lead-based glasses, which don't require as high of process temperatures. Otherwise, we would be completely happy using a soda-lime glass like Bullseye, which has a complete line of colors available. This is all part of the trade off on which type of glass you want to use.

The pâte de verre process temperature is a function of the type of glass you are using and the degree of frit consolidation you are looking for. The degree of consolidation will depend on the kind of surface texture you are looking for—powdery to fully dense. Keep in mind that what we are actually doing here is fusing, but now in three dimensions. The full fuse temperature is the same as that you would use in ordinary flat fusing, it's just that the process times are longer. Also if you are trying to minimize any air in your work, you may want to go slightly hotter than the full fusing temperature.

Donna Milliron tells us that she routinely takes her castings to 100°F over full fuse to get complete consolidation of her work. Other artists, such as Mary Francis Wawrytko, prefer to hold the casting at lower temperatures, she goes to around 1400°F, for much longer periods of time. She does this because the mold remains stronger at this temperature and is less likely to crack on her. She relies on the longer time, sometimes up to a day to allow proper glass flow to fill her mold.

You will have to play with this tradeoff to see what works best for you. We tend to use higher temperatures because we are impatient. Table 12 suggests processing temperatures for some of the glasses that you might use in kiln casting. In normal flat fusing, you only require about a 15-minute soak at temperature. Kiln casting in a mold on the other hand, requires soak times on the order of a few hours. It may be even longer if you are trading off temperature for time.

The kiln casting process temperature that you choose depends upon what you are trying to do at the moment.
- If you are trying to get a full-density casting in a fully contained mold, you would process in the range of the full fusing point.
- If you are trying to get a near full-density casting, while at the same time reducing frit movement, you might process at or just above the tack fusing point for your glass.
- If you are trying to just stick your frit together with absolute minimal movement of the frit, you would process it at or just above the softening point.

The lower the process temperature, the longer the process time will have to be to get flow. For gently sloping objects such as bowls, it helps to consolidate the frit first before proceeding on in temperature to generate a fully fused smooth shiny outer surface. For hollow objects, do not go any higher than the tack fusing point without packing the inside of your mold or you will get gross movement of the frit.

Besides the effect of temperature and time on the consolidation of frit, pressure also plays a role. You can squeeze things together much faster and tighter if you push on it. This is what weights are good for. Use them to push mold pieces together during a firing as the frit gets soft as we did for press molds.

Those of you trying to get very clear castings by melting from a crucible into a mold may have to process at considerably higher temperatures than those stated in Table 12. As much as 300 to 400°F higher may be required to get the glass to the point where it will flow properly and air bubbles to be eliminated.

So here you are holding at the process temperature with great anticipation, waiting for the process to take place. This is only slightly more exciting than watching grass grow. How do you know when it is happening? You just open the door or lift the lid and take a peek. The sharp edges of the frit granules will start to round out and the whole thing will begin to look like orange sushi.

Be sure to wear your protective glasses. Not so much because something will jump out of the kiln and bite or burn you, but because the glow from the kiln itself can be very damaging to those sensitive eyes. There is a lot of invisible infrared and

Table 12. Kiln casting process temperatures for various glasses.

Glass type	Full Fusing Point (°F)	Tack Fusing Point (°F)	Softening Point (°F)	Annealing Point (°F)	Strain Point (°F)
Bullseye					
Transparents	1530	1430	1250	990	920
Opals	1550	1450	1270	935	865
Gold-pink	1450	1350	1180	882	820
Desag GNA	1575	1475	1325	960	800
Wasser	1500	1400	1250	950	650
Uroboros	1450	1350	1225	1000	800
Spectrum	1425	1325	1250	950	700
Plate glass	1450	1350	1275	1150	850
Pemco PB 83	1175	1050	950	775	650
Lenox crystal	1350	1300	1200	826	750
Kugler				875	750

© 2000 James Kervin and Dan Fenton

Kiln Procedures 127

ultraviolet radiation coming out of the kiln that will cook your eyeballs if you give it a chance. Imagine having to take glaucoma medicine for the rest of your life.

But you say that you are afraid to watch the show because your brand new kiln came with warning labels all over it saying "DON'T OPEN THIS KILN WHILE IN OPERATION!!" How else are you ever going to know what's going on inside that hot box? It's almost required to occasionally visually monitor the big event because we're sure no one sold you a program. After all, kiln casting is a spectator sport for the tragically creative and sometimes even requires active participation of adding glass to ensure a proper outcome.

Figure 124. Hungry molds begging for more frit.

You can meet the main intent of all those warnings even if you can't meet the letter of them—by turning off the kiln before you look or reach into it. This will help prevent you from electrocuting yourself. When you do so, you will also want to protect your hands, arms and body because they can be cooked also. Wear heavy cotton clothing, long-sleeve shirts and gloves. For a top loading kiln, it is suggested that you make a hook out of some re-bar to open the lid so as to not expose yourself too much to the heat that will be rising out of the kiln while you hold the handle.

Anyway, let's get back to what's happening. In nprmal fusing, the glass only has to puddle out onto the kiln shelf once it gets soft. In kiln casting, the stuff has some traveling to do. It has to flow down into all the nooks and crannies inside the mold. Depending on the size of these passages and their depth, this journey may take anywhere from an hour to a full working day. So if you don't want to be chained to the stool in front of your kiln monitoring the infinite range switches, it is a good idea to consider getting a programmable electronic controller connected to a thermocouple as will be discussed later.

After you have practiced kiln casting for a while, you may find that a given size mold filled with a particular type glass will fuse to a full density casting at full fuse temperature over a set period of time. So you can just adjust your set point controller to that process temperature and come back to check on progress after that period of time. If it needs more time, then give it more until you have determined that the process is complete.

In most kiln casting firings, it is almost impossible to know ahead of time exactly how long the process phase will take unless you have done this same piece a number of times before. Usually you will have to visually assess the progress to be sure that it is done.

A wind-up kitchen timer with an alarm to remind you to check the kiln is a good idea if you are as drifty or busy as we are. (We are not going to tell you which.) If you have a big studio or if you are hard-of-hearing, go to a photography store and purchase a GAR-LAB darkroom timer. The buzzer on that thing will wake the dead, much less the occasional napping glass artist. The noise is so disturbing that you may even develop a better biological sense of timing just to avoid hearing it. Either that or you may develop some other sort of Pavlovian response and start salivating or something.

Sometimes even though you have visually monitored the flow of the frit down into the mold you can not be sure that it has fully consolidated deep down inside there. If you cut the process time too short, you may notice a gradient in the degree of fusion in a piece. The section nearest the sprue will be more translucent and smoother while that deeper inside the mold will be more opalescent and rougher. If so, consider spending a little more time at process temperature on your next firing.

If you have a set point or programmable controller, you don't have to worry too much about having to get to the kiln right away to prevent over-cooking the glass. Although if your glass is susceptible to devitrification and some are much more susceptible than others, you don't want to spend much more time at temperature then you need to. Devitrification is the growth of an orderly crystalline structure that usually occurs at the surface of the glass and gives the glass a scummy look that obstructs the passage of light through it. Usually all that will happen if you stay longer at process temperature is that the mold will just fill a little better but devitrification can always creep up on you.

© 2000 James Kervin and Dan Fenton

If you have to worry about anything, it's mostly about the mold strength. If it is going to fail at the process temperature it will usually happen in the first hour or so.

As you hang at process temperature, you should eventually notice that the glass frit is sinking down out of the reservoir and into the mold cavity. This will usually be apparent within an hour at process temperature. If that happens, what's a mother to do? You may have to refill the reservoir with more frit to insure that the cavity inside the mold completely fills.

Figure 125. Refilling a mold during a firing.

It helps to have marked your molds or to have made a map of mold locations in the kiln to know which mold gets refilled with which color frit. At temperature, they all look the same — like big glowing blocks.

Depending upon how big of a reservoir you put on the mold, topping off the mold may have to be done a number of times. Some of the more complex inner shapes may be like hungry dogs, eating everything you give them and more. Sometimes the frit will sink down at an almost alarming rate. Unless you see a big glowing mass of glass flowing from beneath the mold (in which case you have one sick puppy and don't need to feed it any more), you just keep feeding him.

Feeding the mold uses the procedure we explained earlier for a cured mold to which you were adding glass directly after curing by using some sort of scoop on the end of a long handle. Figure 125 shows a closeup of a mold being refilled with frit using a metal ladle that has been extended with stained glass rebar. It is fairly easily done if you can remember which color goes where. Otherwise consult your map.

This is also the time to take the usual precaution of turning off the power before going into the kiln. We can not emphasize this point enough. **Always turn off the power before reaching into a kiln.** The kiln is thermally hot with no question but it is the electrically hot that can kill you. That is one mistake you will only make once. Also don't forget to turn the power back on again when you're done or the project will not get finished.

After you are finally satisfied that the mold is completely full and the glass is properly fused, wait and give it another hour or two just to be sure that it fills fully. This is because the inside of the mold will be lagging behind in temperature from the outside. You can then declare the process phase officially over and proceed on to the next phase of the firing.

Cool down phase

The most important aspect of cooling is going slow through the non-brittle temperature regime also known as the annealing range. Annealing, as was discussed earlier, is the process of raising a material to a high temperature where the molecules become mobile enough to relieve internal stress and then slowly cooling the material to minimize any new stress build up. In this case, we are already at the high process temperature, so all that we are really trying to do is minimize stress induced during cooling as we proceed through the annealing range.

In **segment 9**, the first step of the cooling phase, as it is in fusing, is to crash cool the kiln temperature as much as possible from the process temperature (which is in the lower end of the glass's fluid regime) down through the flexible range to the upper end of the non-brittle solid regime at about 1000°F. You try to crash such that the kiln ends up just above the annealing point. This crash is done to try and chill the amorphous glass liquid into a solid as quickly as possible.

Reducing the cooling time spent above 1000°F helps minimize the odds of developing devitrification and after all any time between the end of process phase and the start of annealing is just wasted time. If you are operating your kiln with a programmable controller, you can probably get by without manually crashing the kiln, but as a general principle we recommend always crashing. Besides it makes great party chatter. "What did you do today? Oh, I crashed my kiln." We find that it usually takes between a half-hour to an hour to crash a medium sized kiln.

We manually crash through the flexible regime by venting the kiln. You can do this by opening the kiln door fairly wide and closely watching the temperature drop on the pyrometer. As it comes down to the annealing point, we close the kiln and let the temperature within the kiln equilibrate. At this point, some glass artists will repeat the crash process a couple of more times until the kiln stabilizes about at the desired temperature. Others leave the kiln vented a few inches after the first crash cycle and allow it to crash the rest of the way more gradually.

We have found through experience that rapidly crashing the kiln in either of these ways can be hard on molds. You can hear them plink and chink as they crack before your eyes. Although this may not result in gross flaws, it can result in fine cracks in the mold that can mar your final casting. For this reason, many kiln casting artists just let the kiln crash as fast as it can under no power with the lid closed. This technique is much less likely to result in mold cracks. This is also the technique that you would use if your kiln is cruising along unattended under control of a programmable controller.

In either case, you should realize that in kiln casting the actual temperature of the glass is probably lagging well behind that of the kiln air temperature because of the thermal inertial of the mold. For this reason, in **segment 10** you will want to dwell for awhile at the crash temperature to allow the mold and glass to equilibrate. Remember it is the size of the thermal gradients during annealing that determine the amount of permanent residual stress in a piece. So you want to equilibrate the piece as much as possible before proceeding on.

The length of the dwell in segment 10 is determined by the mass of the glass and by the mass of the investment material (how well it insulates the glass). In addition, the composition of the glass comes into play to some extent. The common fusing rule of thumb for soda-lime glasses is that 15 minutes of soak is required for every ¼" of glass thickness if cooled on all sides without any mold. Where the glass is cooled on a mullite clay kiln shelf, which is highly refractory (holds heat and releases it slowly), the time at the annealing point should be about four times as long. This means that 1 hour of soak is required for every ¼" of thickness to account for the mass of the glass.

We use this same rule of thumb in kiln casting to calculate the extent of the dwell required to equilibrate the glass. This may sound extreme, but keep in mind that the kiln shelf is cooling through the glass. In addition you have to account for the insulative effect of the mold. Plaster-based investment molds hold in heat, but not nearly as much as the dense mullite clay. You should allow 1 hour per ½ inch of mold thickness to account for the time required for heat to flow out through the mold. If you have crashed more gracefully with a closed kiln this additional time can be reduced by ½ to ¼ of this value depending upon how fast your kiln cools; ¼ for most firebrick insulated kilns and ½ for fiber insulated ones.

These two components of the dwell, time for glass and time for mold, are summed to determine a value for the total annealing dwell. Thus if you have two-inch thick casting in a one-inch thick investment mold, you should allow between two and eight hours for the glass and two hours for the mold.

If you have trouble with cracking of castings or for much larger castings, increase the annealing time dwell. After the mold has equilibrated, you may want to add some fiber paper or blanket over the top of open face molds to make heat transfer from the casting be more uniform during high temperature cooling. It will slow loss from the exposed glass surface and make it closer to that which occurs through the mold walls. For small pieces, the soak time can be less. We have found a half-hour per quarter-inch of fused thickness will work.

In **segment 11**, it is now time to slowly cool the glass through the annealing range. We have soaked at the crash temperature for the appropriate amount of time for the glass to equilibrate with the mold. In cooling through the annealing range, we are concerned with controlled contraction of the glass. As the glass loses heat through its outer surface, this area contracts first. The relative contraction between the outer surface and the inner glass is directly proportional to their temperature differences. Therefore, we are trying to minimize temperature differences.

If the glass is cooled too quickly, the outer layer of glass will exert compressive stress on the still flexible glass beneath it. This glass then moves around to minimize the current stress. As the hot interior starts to cool off, it wants to shrink in a different manner but is held in place by the cooler glass shell around it. This causes permanent stress to be locked into the bulk glass. If the glass returns back to room temperature with a grumpy thermal memory, it is more likely to break from temperature changes at a later time from warming by sunlight or hot water.

Thus controlled cooling through the annealing zone is really critical to the ultimate longevity of our work—assuming of course that it makes it through your critical artistic review. This region of the firing is also sometimes referred to as the high temperature cooling zone. It is important that the temperature is reduced slowly through this zone, otherwise the work will not be completely annealed.

This is where, in doing large works, that it really pays to have a programmable controller. If you are firing down manually, make sure that you do not fall out of the zone too quickly or you will have to reheat the piece back up to the annealing point and start over. There is no credit for partial time served.

Determining the annealing schedule for solid glass invested in a mold can be a bit tricky but anything that is not a mystery is really just guesswork. Here the thickness of the glass is the most important consideration, along with its mass. The larger the work, the longer the required annealing times. The refractory character of the investment material, as explained earlier, also influences the annealing schedule. The heat has to pass through the investment material to escape and this can make the effective annealing times as much as four or more times longer than if the glass were cooling without a mold. We usually account for this by longer dwells at both ends of the high temperature cooling stage.

If you happen to be lazy and haven't determined the exact thermal properties of your glass by doing a slump test, there is always the "best guess" method. This is where you cool slowly through the zone where the annealing range is expected. If you figure that the annealing point is at 950°F, then "crash soak" in segment 10 to at lowest 980°F. We usually use 1000°F because it is easier to remember and calculate. This soak at just above the estimated annealing point remember is only to minimize temperature gradients within the glass and the mold before continuing on.

It is the high temperature cooling phase of segment 11 which follows that determines whether you will have any permanent residual stress. This consists of an approximate 150°F drop in temperature from the annealing point to the strain point for soda-lime glasses. To be sure we have completely annealed our work we usually go down in a slow controlled ramp to about 700°F.

In **segment 12**, we soak again once we have reached the strain point. This is to let the heat in the glass pass out through the mold and equilibrate at this temperature. Here you should only have to soak for about ½ the crash soak time in segment 10 since the descent through the annealing zone was much slower.

Cooling rates, as was explained before, are controlled mainly by glass thickness. Refer back to Table 4 in the chapter on glass to determine an appropriate high temperature cooling rate for your firing. (You may also want to review the discussion there to refresh yourself on how those cooling rates were derived and how they may change some with different glasses.)

With small pieces, several cubic inches or less in size, the mold can serve as an advantage since it slows the heat loss down through the annealing range. In such cases, you may even be able to turn off firebrick insulated kilns after the process phase and allow the piece to drift down naturally through the annealing range.

As your work gets larger, controlled annealing becomes a definite necessity. This is because, as you should remember, annealing times are calculated based on the thickest part of the glass to be safe. For really large work like that done by Linda Ethier or David Ruth, the annealing cycle can take as long as a month.

Here besides size, the shape of the work is also a factor in how much annealing is required. A regular shape, like a square or a sphere, will require much less annealing time than one with the same maximum thickness but having a lot of thickness variation. This type of work is more likely to develop large temperature variations and thus stress. Residual stress seems to concentrate at the thin points causing the piece to snap apart at this cross section.

In **segment 13**, once safely equilibrated below the strain point, we transition to the low temperature cooling portion of the firing. Here the cooling rate is not as critical as in the high temperature cooling range since all permanent stress is already frozen into the glass. There is no changing the glass's thermal memory at this point, and all temporary strain caused by thermal gradients in this stage will

Table 13. Summary of a typical firing schedule.

Segment	Description	Temp Change (°F)	Rate Range (°F/hr/in)	Fast Time (hr)	Slow Time (hr)
1	Low temperature heating ramp	RT-225	50-300	½	3
2	Physical water removal dwell	225		1	2
3	Low temperature heating ramp	225-350	50-300	½	2½
4	Chemical water removal dwell	350		½	1½
5	High temperature heating ramp	350-1200	100-600	1½	8½
6	Organic burnout dwell	1200		1	2
7	High temperature heating ramp	1200-1550	100-600	½	4
8	Process temperature dwell	1550		3	4
9	Kiln crash ramp	1550-1000		½	2
10	Crash equilibration dwell	1000		1½	5
11	"Annealing" cooling ramp	1000-700	25-150	2	12
12	Strain point dwell	700		¾	2
13	Low temperature cooling ramp	700-RT	50-300	2	12
Total				15¼	60½

RT is short for room temperature.

dissipate as soon as the glass stabilizes back at room temperature.

Of course if you go too fast, the temporary stress can still build up to high enough levels to crack or thermal shock your work. This is not as likely in kiln casting as it is with fusing because the mold usually slows down heat loss enough; unless of course it's an open-face mold. So many times, you can just turn the kiln off after the high temperature cooling phase and allow it to drift back down to room temperature on its own.

To help make an open face mold cool more uniformly, you again might want to lay some fiber blanket over the top of the mold. Our firebrick kilns take about 8 hours to cool back to about 200°F from the strain point. It is usually only with larger works that you have to fire down through this phase since they are more susceptible to thermal shock. Here you can get by with doubling the cooling rate over what you used in segment 11.

After the kiln has cooled, allow some time to ensure that the mold and the glass have really returned to room temperature. Since the glass cools through the mold, it could still be warm inside. In other words, even though you can lift up the mold with your bare hands, the glass on the inside can be hot enough to burn your fingers. It is better to allow some extra time for the glass to cool than to chance having to go through all this a second time to make a new piece.

So how much time do all these steps add up to anyway. Table 13 lists all thirteen firing schedule steps that we have discussed in this chapter, gives some suggested ranges of both temperature change rates for the heating and cooling as well as dwell durations in each segment. This table is calculated for small projects on the order of an inch or so in thickness encased in an inch thick mold.

You will have to decide whether you want to be on the fast or the slow side depending upon the complexity of your work and how aggressive you are.

Kiln controllers

As you can see, the process of kiln casting of glass can be rather lengthy. The total kiln time itself may vary between hours and days depending on the size and intricacy of your work. To achieve optimum results, your kiln will need close control. So if you do not want to be a slave to your work, you will want to consider getting some sort of electronic controller for your kiln. We will finish out this chapter with a discussion of kiln temperature control and measurement.

The standard control mechanism on most kilns are infinite range switches. They are the dials that have low, high, and a bunch of other settings in between on them. They are seen on the left of the kiln in Figure 126. They control the amount of time that electricity is flowing through the wires of your kiln elements.

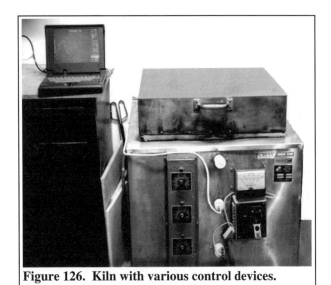

Figure 126. Kiln with various control devices.

Set point controllers

The simplest form of intelligent controller is a set point controller. It can hold a kiln at a given temperature indefinitely. It will keep your kiln within a few degrees of your set temperature by turning the power off when it is too high and turning it back on when the temperature drops too low.

This type of controller will not regulate rates of temperature rise or fall. You have to do that by adjusting the infinite range control switches. They act as a governor for your kiln, setting the maximum operating power fraction to help keep the kiln from getting out of control. In this way, they can be used to act as a manual override over any electronic controller.

So with a set point controller, you can set your infinite range control switches for the approximate rate of temperature rise that you want and the set point controller for the final temperature. Then when the kiln reaches the set temperature, it will hold at that temperature. If already at that temperature, it will stay there by allowing the elements to come on and off as necessary to remain there. Set point controllers are not nearly as expensive as they used to be and many kiln manufacturers now include them as a standard option on new kilns.

Kiln sitters

We would caution against using "kiln sitters" to control glass firings unless you think it is possible that you might forget about the firing and have no other way to automatically turn it off. A kiln sitter is a more or less idiot-proof device that shuts off the kiln after a given amount of heat work. It works by

They work in a similar manner to the thermostat for your house. Inside they have a bimetallic strip (two metal strips welded together) that bends as it heats up from the current flowing through it. So current will flow through it for a while until the strip heats up enough to bend and break electrical contact. Then as the strip cools, it straightens out enough to again make contact. By rotating the dial on the switch, you control how far the strip has to bend before it breaks contact and thus the fraction of time that the kiln is on.

You might think that this would be a good way to control the temperature of your kiln. Unfortunately it isn't. It does not read what is happening in the kiln and must be manually adjusted. You have to work a lot with your kiln to get a feel for how the different infinity switch settings control its rate of temperature rise.

Figure 127 illustrates the rate of temperature rise for different infinite range control switch settings on one of our kilns. From this figure, you would think it possible to choose a setting that plateaus out at the temperature that you want. There are two problems with this approach. First, the rates and final plateau temperature vary some with the mass of the material in the kiln. Second, you usually want to get done as fast as possible because kiln time consumes electricity which isn't free.

Wanting to get done as fast as possible means you want to raise temperature as quickly as possible regardless of final plateau temperature. For that reason, you may want to consider some sort of electronic controller.

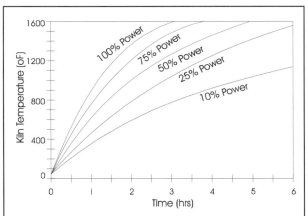

Figure 127. Rate of temperature rise at different infinity switch positions.

having a small weight in the form of a lever arm resting on a cone placed in a horizontal position. The weight of the arm exerts a force on the cone that causes the cone to slump over time. The cone measures heat work and slumps when that has been achieved.

This type of controller has the problem that the glass may need its heat work at a specific temperature where its viscosity is lower while the cone may register what it believes is the approximate heat work at a lower temperature with longer time.

This type of control is also not appropriate for a situation where we want to go to peak temperature for a while and then ramp back down in a controlled manner since the kiln sitter will turn the kiln off. As you can see in the lower right side of Figure 128, Jim has one still installed on his work horse kiln. Here he uses it mainly as a backup timer for ending a kiln run in case his computer based controller develops bugs. Actually the kiln came that way and he was really just too lazy to remove it.

Programmable controllers

Programmable controllers can control just about every aspect of your firing — ramps, dwells, etc. They vary in how much power they are rated for, the number of program segments a firing program can have, how many programs they can store, how many kilns they can control, and how they control ramp segments. Let's discuss each of these points in greater detail.

Most kilns will run either on normal household power of 110 volts or on large appliance power of 220 volts. They can not be switched from one to the other. At the very least, the power relay of the controller, the device that turns the power on and off to the kiln at the command of the controller, will have to be changed to a new one of the proper voltage and power rating to match the kiln. By power, we mean that besides voltage, the relay also has to be rated to carry the current required by the kiln to heat the elements. The product of current (amps) and voltage is power and is measured in watts. So when you go looking for a kiln controller be sure that it is rated for the current and voltage required for your kiln.

The next features under consideration have to do with how versatile you will find your controller. The first of these, the number of segments allowed in a program, is the number of ramps and dwells that you will be able to have in a firing. As we saw in this chapter, a typical pâte de verre firing may have as many as thirteen segments or more. Some controllers only offer a limited number of control segments and therefore are not that useful for kiln casting.

If you do more than one kind of firing with your kiln, it would be nice not to have to completely reprogram the controller every time you want to do a different operation. This is where the feature of being able to store more than one program in the controller comes in handy. You could have programs already entered for a number of different operations (wax melt out, mold burn out, initial consolidation, final casting, fusing, slumping, etc.) and just punch a button for whichever one you wanted at the time.

As your studio grows, the number of kilns that you have may also grow. You might start doing techniques that require more that one kiln to be operating at a time—like pouring molten glass from a crucible heated in one kiln into a preheated mold in another kiln. Alternatively, just the long firing times of large pâte de verre pieces may convince you that you need more than one kiln.

Lastly let's discuss the two different methods that manufacturers have implemented for ramp segment control. They are end point and rate control. In **end point control**, you program the time and temperature for each change in your program. Thus for a simple program where you go from room temperature to 1000°F at 200°F/hr, hold for an hour and cool at 200°F/hr back to room temperature; you would have to program four points.
- First you program the starting point as 0 time 0°F (easy math temperature or very cold room).
- Next you program the end point of the first ramp of 5 hours (1000°F/200°F/hr = 5 hr) or 300 minutes and end point temperature of 1000°F, the controller will recalculate the rate and control appropriately.
- The third point will be 6 hours (5+1) or 360 minutes and end point temperature of 1000°F.
- The last point will be 11 hours (5+1+5) or 660 minutes and an end point temperature of 0°F.

In **rate control**, you program the temperature rate and the length of the segment. So for the same program you would program three segments.
- The first segment would be a rate of 200°F/hr for 5 hours or 300 minutes.
- The second segment would be 0°F/hr for 1 hour or 60 minutes.

- The last segment would be -200°F/hr for 5 hours.

Either of these techniques works just fine. End point control is the most common type on the market. Dan has a Digital controller of this type. Some people may prefer thinking in terms of allowable rates and for that reason might prefer rate control.

Most of the controllers have a keypad and a small LED or liquid crystal numeric display where you step through the program, one segment at a time. Because of screen limitations, they may also require that you learn some sort of a simple program language to communicate with the controller. There is also a PC run controller — Kilntrol IV — that allows you to visually program your kiln with a mouse on a computer screen. It displays where you are in the program but this controller requires a 386 or newer PC. This is the type that Jim uses and is the reason that he has the laptop computer out in the left of Figure 126. It is also an end-point controller.

Whichever kiln controller you purchase, it won't be long before you become proficient with it and wonder how you ever got along without it. So why wait? Go get one.

Pyrometer calibration

To accurately control your kiln and determine glass properties, it is required that you have a well calibrated temperature measurement system as seen in the center of Figure 126. The preferred temperature measurement system is a pyrometer. A pyrometer (translates literally to mean fire meter) uses a thermocouple made from wires of two dissimilar metals that is inserted into the kiln to monitor its temperature. Heating the welded junction between the two wires causes the metal in one of the wires to pull electrons from the other wire.

This physical effect, called the Seebeck effect, causes electricity to flow around the electric circuit from the thermocouple to the pyrometer. It is actually a quite common effect between metals, but certain metals produce much higher voltage potentials than others and are therefore used for this purpose. Some specially developed alloys, like chromel, alumel, and constantin have been developed to enhance the Seebeck effect and make thermocouples so sensitive that they can measure the temperature increase from holding them in your hands.

A pyrometer actually measures the amount of current that is flowing through it, but since the current is directly proportional to the temperature of the thermocouple, the readout is calibrated in units of temperature. The markings on the meter are equally spaced but this is really only an approximation. The Seebeck effect is really non-linear so the readout is only accurate over a narrow temperature range.

Another thing that you should be aware of is that not all thermocouples are created equal. They vary in materials used, wire diameter and sheathing technique. Table 14 lists some thermocouple types, the metals from which they are made and their applicable temperature range. They each provide different electromotive force potentials for a given temperature and thus shifting from one to another would require recalibrating your pyrometer. Figure 128 shows a sheathed thermocouple on the to and a unsheathed one beneath.

Wire diameter and sheathing technique both affect the responsiveness of your thermocouple. The thicker the wire or the more sheathing insulating the welded junction, the slower the thermocouple will respond to temperature changes. But at the same time, the thinner the wire and sheathing, the sooner the thermocouple will burn out. The function of the sheathing is to protect the thermocouple from attack by the kiln's atmosphere especially a reducing one, which can eat away at the metal. Stainless steel sheaths are usually used but we have also seen good quality thermocouples where fast response is desired use quartz as sheathing.

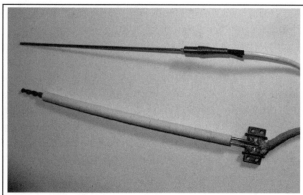

Figure 128. Two types of thermocouples.

Table 14. Some common thermocouple types.

Thermocouple Type	Metal pair	Electromotive force (mV/°C)	Temperature range (°C)
T	Copper Constantan	0.0509	-200 to 300
J	Iron Constantan	0.0529	-200 to 1100
K	Chromel Constantan	0.0720	0 to 1100
E	Chromel Alumel	0.0407	-200 to 1200
S	Platinum Platinum-Rhodium 10	0.0105	0 to 1450
R	Platinum Platinum-Rhodium 13	0.0117	0 to 1450

Since the most critical temperature region for kiln casting is in the annealing range around 1000°F, we usually calibrate our pyrometer for this region. To do this, we place a large # 022 cone in a wire stand near the thermocouple junction. This cone will indicate a temperature of 1090°F. As you may know, cones actually indicate heat work, the integrated time/temperature history of your firing, and not the exact temperature of the cone. But by controlling the temperature rise rate of your kiln to one of approximately 270°F/hr as used in calibration of the cone material, it will give a fairly accurate indication of 1090°F.

The cone is set in the stand at an angle of eight degrees. When the cone slumps to "wicket," which is where the tip of the cone bends over and touches the deck, you have achieved the calibrated heat work. You might also want to place a # 021 cone in there with it in case there is an over firing. This cone will indicate a temperature of 1130°F.

Ceramic artists will usually use three cones when doing a firing. One rated for just over the firing temperature and one rated for just under it. The third is for the desired heat work. The problem here is that a # 022 cone is the lowest temperature cone currently available and is closest to the temperature range in which we are interested.

Other more accurate tools that can be used in calibrating your pyrometer are chemical temperature indicators marketed by Omega Engineering, Inc. They have tablets, labels, and lacquers that change color or melt at different temperatures, many of which are lower than a # 022 cone. There are indicators for over 100 different temperatures ranging from 100°F to 2500°F some of which may be more appropriate for calibration in the annealing range. We have used a 950°F lacquer that works well. In fact it works almost too well because if you aren't watching closely it will be there one minute and gone the next. One call to Omega's economically correct phone number will get you 25 pounds of free catalogs without the hassle of a resale number or the governor's signature.

Once you have determined your pyrometer reading at a precise test temperature using cones or some other means, you can adjust the pyrometer or controller readings, if necessary, to indicate the correct temperature reading. On the common needle-type pyrometers, there is a set screw on the front of the meter to adjust the reading. Turning the screw gives you access to about 50°F of adjustment one way or the other.

You will find though that adjusting it in one range will not assure a correct temperature reading in another range. When calibrated to 950°F, it may be as much as 50°F off at 1550°F. This is because, as was mentioned, the temperature induced voltage of the thermocouple is not necessarily linear as interpolated by the meter. This problem is not the case with a digital electronic readout pyrometer which is a good thing since it is much more difficult to determine which series of screws to turn. Of course reading the directions might help but that takes the fun out of it. All those adjustments compensate for the non-linearity of the thermocouple response and once calibrated a digital pyrometer will indicate temperature accurately over the complete range of interest.

"Promises" 1996
Lost wax kiln cast crystal
Size: 15½" x 10¾" x 10¾"
Artist: Kathleen Stevens
Photo by: Arthur Probst
Shown at: The Divine Gallery

Finishing Techniques

The final step in any kiln casting project is cleaning up your glass casting. How much finishing work you have to do may vary a lot. Some of the things that you may have to do to finish a kiln casting will include:
- You may have to trim off gating or sprues.
- There may be flashing from mold joints or cracks that has to be removed.
- Parts may need to be further shaped to fit them together.
- The bottom or base may need flattening to get the casting to stand correctly.
- Activated mold or clay residues may need to be removed.
- The surface of a part may need polishing to make it shine.
- You may want to add a thin layer of low melting point enamels to shine it up

These are just a few of the things that may need to be done to finish a kiln casting and make it into something that really captures a person's eye. To do them will require tools beyond what we have discussed to this point. It will require an array of abrasive cutting and grinding equipment.

Generally the finishing steps would be in this order:
1. You first cut off any extra glass such as sprues or flashing with saws.
2. You next rough grind down any unintentional raised areas using coarse grit grinders or sanders.
3. You then smooth out rough ground areas with further cold working using medium grit grinders, stones, or sanders.

If you are satisfied with a matte finish, then you, as many artists do, would stop at this point. If you want a shinier finish, you would continue onward.

4. You would next semi-polish the piece using fine grit sanders, wooden wheels, and brushing wheels.
5. You would then do a final polish of the piece on a buffing wheel.

Most of these steps fall into the realm of what is commonly called cold working of glass. Very little of this information is readily available in the literature, so we hope that this chapter will explain some of its mysteries.

Diamond embedded tools

Most of the abrasive tools, we use in finishing glass castings are diamond embedded tools. Diamond-embedded tools are used to cut glass because diamond, being one of hardest materials known can easily cut away at glass. Its particles are sharp and so much stronger that the glass is readily scratched away.

You may say to yourselves, "Diamonds, who could afford those?" After all, we all know how much engagement rings cost. If you believe your neighborhood jeweler, they should be at least a couple months' salary. So how could we ever afford to use them for cutting glass. Well wipe that image out of your head. The kinds of diamonds that we will be using are industrial diamonds that are not all that expensive. They are very small and man-made.

They are not sized according to carats but instead are sized by grit or mesh size. They are measured by passing the particles through wire mesh or screens just as frit is. The mesh size of the screen is defined by the number of openings per inch it has—the larger the mesh number, the finer the grit. A list of mesh sizes and their corresponding dimensions were given in Table 2 in the chapter on glass.

The mesh size of diamond that you want depends upon what you are doing and what kind of finish you are trying to achieve. Large grit sizes, those with smaller mesh numbers, will leave a rough surface but will really cut through a piece of glass. They are used for cutting and rough shaping. For this, you would want a coarse grit of about 80 mesh or larger.

The next operation would be grinding or smoothing. Here you may again start with an 80-mesh tool to cut quickly to the rough shape. But you would then probably follow that up with use of a medium grit tool of about 220 mesh, and then by one with a fine grit of about 400 mesh. For a surface that will show, you would finish with polishing tools using something like cerium oxide polishing compound. Some artists may prefer using finer grit in the 500 to 600 mesh range before moving on to polishing.

The diamonds on your tools are usually bound in a matrix to hold them in place on the wheel. They are not that cheap that we can afford to keep replacing them all the time. The matrix will be in the form of a softer material like an epoxy adhesive or a soft metal like copper. If a diamond embedded tool overheats, the matrix can get soft and loose diamonds. For this reason, we always use a lubricant with them even if it is only water. Usually though you will use a diamond cutting lubricant which are water-soluble oils.

Besides diamonds, some tools may also use silicone carbide. It is another good choice for a grinding material first because it is also one of the hardest materials known. In addition, it has the desirable property that when a silicon carbide crystal breaks, it forms another identical sharp fracture edge regardless of particle size. Thus, the grit continues to stay sharp as it breaks down into smaller and smaller pieces. These two properties combine to make it an excellent grinding agent, second only to diamonds in cutting capability. It is usually used on solid metal wheel machines where it is applied as a powder slurry of grit in water.

Let's next discuss the different types of finishing tools that we use to cold work kiln castings and how to use them.

Cutting a casting

As mentioned above, there will be times when you are faced with the necessity of having to remove excess glass from your castings. This can be in the form of gating, flashing, mold flaws, or simply burrs. The tools that we would use for this purpose are diamond-embedded glass saws.

Types of glass saws

Glass saws use diamond embedded blades to grind their way through a piece of glass. They come in a number of configurations: band saws, wire saws, cutoff saws, table saws, powered circle saws, and handsaws. In all cases, the saw will use a cutting fluid to keep glass dust out of the air and to keep the blade cool. You want to keep the blade cool for two reasons. First, so that the diamonds are not melted out of the blade's binding matrix. Second so localized heating of the glass does not crack it. Other sources for saws are tile saws and lapidary saws.

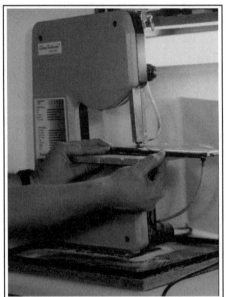
Figure 129. Small diamond band saw.

Band and wire saws

A band saw has a blade made from a continuous loop of metal band or wire that runs between two drive pulleys through a table that has a slot in it. A guide supports the back of the blade. It reacts the force from pushing the glass against blade. Water is fed directly onto the blade from a reservoir above the blade or by passing the blade through a water-filled reservoir.
- If the water is fed from above, you will have to be sure that the feed reservoir is full and the feed valve is adjusted properly.
- If the blade picks up the water from a reservoir below the table, then you just need to make sure that the reservoir is filled at least to the level of the blade.

To cut glass off of a casting on a diamond band saw, slowly push it up against the front of the blade, and feed it in as the glass gets ground away. One of the disadvantages of a band or wire saw is that the blade tends to wander a little as it cuts through the glass casting leaving a rougher surface than that achieved with the other types of saws. But the flexibility in how you can cut freeform shapes makes them very powerful tools.

Figure 130. A diamond cutoff saw.

Cutoff saws

A cutoff saw has a configuration similar to that of a radial arm saw. With this saw, you pull the spinning disk shaped saw blade toward you and it cuts a slice off of the glass casting that is being held firmly on the table. Cutting fluid feed will be fed either from a reservoir above the blade or from one below the table. The same cutting lubricant adjustment requirements as for the band saws apply. The one in Figure 130 is a MK-101 Pro Tile saw.

Because cutoff saws, like many of the saws that follow, have blades that are large disks, you do not have the flexibility in how you trim off material as you did with a band or wire saw. Here you will have to approximate a smooth curved cut with a series of short straight cuts. Inside curves are especially hard to do.

Any disk saw blade should run true as it spins. You should not see any wobble in the blade. Normal wobble should not exceed more than about 0.0005". If your blade wobbles more than this, then it or the saw arbor may need to be repaired.

Table or trim saws

A table saw, like the one seen in Figure 131, has a configuration where the spindle of the blade is mounted below the table. It is a PLASPLUGS® Power Master™ Diamond Wheel Wet Saw. Table saws such as this will often be cheaper than cutoff saws because of their simpler construction. If you order one from a lapidary house, you may find them referred to as trim saws. They do not have to have a mechanism to move the blade.

Diamond cutoff and diamond table saws can often be purchased more cheaply through tile supply stores than from glass equipment supply ones.

With this type of saw, the blade picks up cutting fluid from a reservoir below the table. The fluid level should be high enough so that the blade is immersed at least three-eighths of an inch deep into it. Any less and the blade will not be properly lubricated. The lubricating solution also washes debris from the blade, keeping the diamonds exposed.

To trim a casting, you push it gently up against the spinning blade. Watch the blade as you make your cut to ensure that you are not deflecting it. This can happen if you are not feeding the casting straight into the blade or by twisting the casting slightly. This can result in a ragged cut and damage to the blade. It can become dished and will have to be sent out to be flattened.

Reversing the blade (taking it off, flipping it, and putting it back on the arbor) periodically will allow it to wear more evenly.

Some people believe that using a cutoff saw is easier than a table saw because they can setup their cut and then just bring the saw blade down into the piece. Others prefer using a table saw because it leaves both hands free. They can then have one hand on either side of the cut to prevent breakage from the glass cracking the last little bit as the section of casting starts to sag.

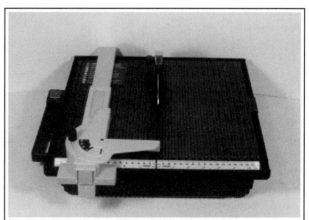

Figure 131. A diamond table saw.

The table for all three of the previous saw configurations will usually have a guide against which to hold the casting and feed it at the desired angle into the blade. This will help stabilize the casting as you feed it into the blade. The angle of the guide can often be adjusted for something other than a perpendicular cut.

Powered handsaws

A powered handsaw, like a circular skill saw is a good tool to use on larger castings where it is not easy to move the casting around like you may want. A number of manufacturers make nice small economical saws that can be used for this purpose. One of these can be seen in Figure 132. It is a Superior Tile Cutter 84-8210N. Powered handsaws have $3^{3/8\text{th}}$ to 4 inch diameter blades, which are cooled by water from a small attached water supply system. Mark Abilgaard says that he uses his Makita handsaw a lot in his work.

These are neat little tools and come in really handy but the same type of cautions that we gave for other disk bladed tools holds for these. Go straight through the casting and don't twist the blade. Keep the blade well lubricated.

Figure 132. A small powered handsaw.

Handsaws

A simple handsaw is a last type of glass saw that you will probably want to have in your arsenal. There may be times when it may be the only tool to be able to reach in where you want to cut. These come in the form of a hacksaw or a coping saw blade that you mount into a frame as seen in Figure 133. You can also just hold the saw blade by hand.

You may hear them referred to as diamond notch saws because they are used to notch plate glass. Obviously they will only be able to saw within the reach of their throat. This will usually be a couple of

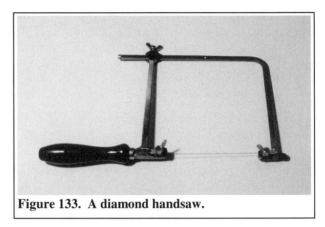

Figure 133. A diamond handsaw.

inches or so. Like band saws, the blades of these saws will tend to wander some as you cut and not give you a completely smooth cut. Lubrication during cutting in this case can be provided by stopping and using a squirt bottle.

Using saws to cut glass

New saw blades should be broken in before being given heavy use. You want to clear away some of the soft matrix from the wheel to expose diamonds. This will allow better lubrication and will increase the cutting speed of the blade. New blades are broken in by cutting a soft abrasive material like a common red building brick.

Occasionally, a diamond blade may get glazed over from glass loading of the blade. This is much more likely if you do not properly lubricate the blade. If this happens you will have to break in the blade again. Make a couple of cuts through a red brick like before to expose diamonds and make sure you have the proper amount of lubricant.

As mentioned previously, you should always use a cutting fluid to lubricate the blade. If you do not, you will severely shorten the life of your blade by causing it to heat up. This will weaken the bonding matrix causing it to lose diamonds faster. You can also heat up the glass enough to cause it to crack from thermal stresses. When the lubricant gets dirty after a lot of use, it is time to clean it out. We usually just dump all the lubricant and scrape all the glass sludge from the tank using a small spatula. We wrap this sludge up in newspaper for disposal.

The other reason you want to lubricate the cut is to keep the glass dust down. The last thing that any of us wants to do is get silicosis from inhaling too much glass dust. Even though the cutting fluid helps keep glass dust down when sawing, it is also recommended that you wear a respirator with a mist cartridge or at the very least a dust filter when using any glass cutting or grinding equipment. This is because small cutting fluid droplets with glass dust in them are thrown into the air and can be inhaled into your lungs. Make sure that your fluid reservoir is full before you start cutting. Be sure to read the equipment instruction manuals for more complete details.

Besides your lungs, you will want to protect your eyes. Wear safety glasses to prevent any chips from getting into them. Do not rub your eyes with your hands until after washing your hands as you will have all sorts of small pieces of glass them.

Since the as-cast edges of many castings can be very sharp, you may want to wear gloves at least until you have trimmed or ground off those sharp edges. Keep a good hold of your piece as you cut it so that it will not go flying. This can be even more of a problem if you are not careful to feed the glass straight into the blade. If you feed the piece crooked, the blade will bind on the glass and try to pull it out of you hands.

We are of the faction that prefers the use of table saws instead of cut-off saws. We find it easier to support the piece on either side of the cut by hand than having to jig it up for support. This is what you should do if you are using a cutoff saw. If you don't jig it, the piece will sag and crack during the last little bit of the cut every time. Clay is the ideal material for making those jigs.

If there is an orientation that provides more support on the table, use it. As an example, cutting the flashing from around a bowl formed in a two-part press mold is much easier when the casting is positioned with the top of the bowl flush on the table unless there is a lot of raised decoration on the side as seen in Figure 134. If you stood it on the bottom of the bowl, it would not be nearly as stable. Remember to cut this flashing off in a large number of straight cuts to approximate a circular cut around the part if you only have a table or cutoff saw. This is one situation where a band saw is really the right tool because you can get into and follow the curve around the piece.

Even with jigging or support, the glass will often chip out on the last little bit of a cut. This happens just from the pressure of pushing the glass against the blade. To help minimize this, go slower, especially near the end of a cut. If you do not want the end of the cut to chip, you could cut a little notch into the backside of the piece before you cut through the piece. This will restrict any chipping from going to the outer surface of the glass because of the much further distance it would have to go.

Many artists will cut a groove all the way around the piece before trying to cut through it just for this purpose. Some will even continually rotate the piece that they are cutting to avoid getting any chipping as the blade breaks out. It can be hard to get this groove in one plane so we usually wait until we get most of the way through the piece before doing this. This makes eyeballing the right position for the groove easier by using the finished portion of the cut to establish a baseline that we extend.

As your pieces get larger, it is harder to get the blade up against the piece to cut off protrusions close to the piece. Instead you can only cut them as close as possible and then grind them the rest of the way down. You may want to leave as much as one sixteenth of an inch to grind off to get a smooth surface.

Grinding a casting

Once you are down to the stage where you need to grind smaller imperfections or the remains of larger ones that you have cut off, you are ready to move on to grinding equipment. There is an even larger variety in grinding equipment than there was in cutting equipment.

In grinding, you may go through several steps. The exact number of which will depend upon the level of surface finish that you are trying to achieve. First, you would rough out the changes you are making using a coarse grit piece of equipment. Grit of 80 to 120 mesh would be about right for this purpose.

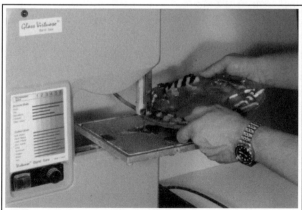

Figure 134. Cutting flashing from press-molded bowl.

Next, you would smooth out the cuts using a little finer grit. We recommend tooling with grit of about 220 to 400 mesh for this step. If this area of the casting is a place that is not going to show, like the bottom of the base, you may stop at this point.

If the area will be visible, you may want to proceed on to polishing steps. How fine you go to depends upon the final finish that you are trying to achieve. Since most pâte de verre has a matte or semi-matte finish, many artists do not go much past about 240 grit. Each of these steps requires slightly different equipment.

Grinding equipment

Figure 135. A large steel disk grinder.

Grinding equipment comes in different shapes and sizes. These different configurations each allow us to do something different. They may for example allow grinding of a different surface geometry. Let's look at some of the different pieces of grinding equipment.

Disk grinders

A disk grinder is great for grinding a flat surface like a base on a casting. They consist of relatively large metal disks that are mounted so that they spin on a vertical axis. If you are looking at lapidary equipment, you may find these tools listed under the name of laps or lapping equipment. They are especially handy when trying to grind a flat surface on a casting. This is something that is hard to do on a wheel.

The disk for the grinder will be mounted to a sturdy square frame with a catch basin around the wheel. The arbor and drive motor are mounted below the basin and the drive shaft will feed up through the basin. The motor will drive the arbor through a v-belt.

The disks for these grinders come in a wide variety of sizes, anywhere from about four to twenty inches in diameter. If you decide to get one, you want the diameter of the disk to be at least an inch or two larger than twice the size of the largest casting you anticipate that you will be grinding. You want your casting to be on one side of the disk or the other. It should not straddle the center.

Most, like the one seen in Figure 135, are just steel disks that use a slurry of silicon carbide grit in water to lubricate and cut the piece. Others, like the one seen in Figure 136, use diamond embedded disks. It is a Denver Glass Machinery DG-18 grinder. This kind will just use plain water with diamond cutting oil to lubricate the disk. They will be run at speeds anywhere from 300 to 1300 rpm depending upon the size and make up of the disk.

In either case, a grit slurry or clean water is feed onto the center of the wheel. From there, it is allowed to flow uniformly out over the wheel to the edge and out onto the inside of the catch pan. For a disk grinder using a slurry cutting solution, there is usually a water source that feeds into a funnel-like arrangement and then onto the disk.

Figure 136. DG-18 diamond disk grinder.
Photo Courtesy of Denver Glass Machinery.

As the water passes through the funnel it goes over a pile of silicon carbide grit. As it flows over the pile, it will pick up some grit into the stream and deposit it onto the disk. After a while this pile will get depleted and will need to be replenished. Just scrape a spatula around the inside of the catch pan to recover some of the grit that will have collected there.

Finishing Techniques   143

When grinding on a disk grinder, do not hold the casting stationary. Instead, glide it back and forth across the cutting surface. This motion will help lubricate the cut and remove any trapped glass particles that you are grinding off. This will prevent them from binding up and causing chipping or scratching of the casting. Sweeping the casting back and forth across the grinding surface as you work it will also allow more even wear of the disk. If you do not do this you can wear grooves into it like on a phonograph record. If any of you do not know what a phonograph record is, then you are way too young or we are way too old.

If you get grooves or waves in your cutting surfaces, you've got problems. It will grind waves into the bottom of your piece that you will have to remove with subsequent steps. The best thing to do is dress up the wheel to smooth it back out into the desired shape. This is done by using a block of abrasive material that you run back and forth across the running tool just like you were supposed to do with your casting. This will grind the grinding surface smooth again.

Whenever using any piece of rotating equipment, either a disk grinder or the spindle mounted wheels, there are some particular steps that need to be followed to prevent injury. The first of these is to prevent getting twisted up into the equipment. Jim has one friend that he works with that had an arm pulled off by a machine shop lathe. Most of our equipment is not that powerful but it could do things like wind up scarves that could strangle you.

So don't take a chance. Avoid wearing loose flowing clothing. Tie back your hair. Don't have any tied belts from aprons in front of you. Roll up your sleeves tightly so that they are out of the way. Don't reach around rotating equipment. Wear goggles to protect your eyes. A plastic facemask might even be better because particles tend to get spread around. Because of this, wearing respirators is also always a good idea.

Spindle grinders

Spindle grinders are mainly used for removing burrs or flashing that is not easily removed with saws. They can also be used to slightly hollow out an area like the base of a casting. They consist of abrasive wheels mounted vertically on a horizontal shaft, or spindle. With spindle grinders, you use the outer edge or rim of the wheel for grinding as compared to the flat side as was used for disk grinders.

The spindles are usually run from a 1725-rpm motor that is of quarter horsepower in strength. These motors are used because they are very common in household appliances and are thus commonly available. The spindle is run by a V-belt from a pulley on the motor to one on the shaft. To change the spindle speed for different operations, the belt is moved from one pulley to another on the spindle.

A spindle will be either single or double. Double spindled units are more common because once you have the motor and the shaft in place it makes economical sense to run more than one spindle and grinding wheel from it. They will be mounted at a comfortable height above the table to allow easy access to the wheel. The mounts for the wheels are designed with threads so that the wheels are tightened on them as they spin.

The motor is mounted such that the direction of the spin of the wheel is down and away from the operator. Because of this, you should always work with your piece on the lower half of the wheel. If you face the wheel from the left side with the hood on the left of the wheel, the area that you should work on the wheel would be between where four and six o'clock would be on a clock.

Figure 137. SB-810 studio glass beveler.
Photo courtesy of Denver Glass Machinery.

Using the area past six is not practical because you do not have good access to it. Using the section of the wheel above this, between twelve and four would be extremely dangerous because the wheel can throw the work back at you. Even when you work in the correct section of the wheel, the piece can still be snatched from your hands, but here it will be thrown down under the wheel and not at you.

Whenever you work on any piece of equipment on a spindle you should always be wary of it grabbing the piece from you. Try to hold the piece so that if this happens it would not pull your fingers with it. This means be careful of rings or bracelets that could hook on to a piece. Hold your piece by the edges so it can be ripped from your hands without hooking anything. Be especially careful when the piece still has sharp edges. Consider wearing gloves at this point.

Depending upon the composition and grit size of the wheel, spindle grinders can be used for grinding, smoothing, or polishing operations. They may use diamond-impregnated, steel, stone-like, wood, cork, or felt wheels. Each different material type or grit of wheel is used for a different stage in the finishing process.

Wheels also come in many shapes. They may be thin and triangularly shaped to reach into narrow grooves or flat and wide for smoothing. Edges also come with rounded bullet-shaped edges. The most common are wheels with simple flat edges.

By positioning the work at different locations or angles to the edge of the wheel, you can often change what kind of cut you make. For example, with a flat wheel you can either make flat cuts on the outer edge of the wheel or you can use the edge to dig concave cavities into the piece. Of course, whatever surface you use must have grit and coolant on it or you can run into problems.

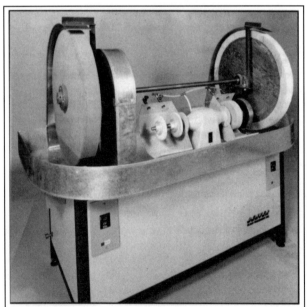

Figure 138. Stone-wheeled spindle grinder on IB-30 Courtesy of Denver Glass Machinery.

Some types of wheels wear more than other kinds. Stone wheels, which are made from soft materials like pumice, are notorious for this. Silicon carbide wheels also have only a soft to a medium bonding matrix so they will wear too. If your wheel starts to develop low spots or grooves then you will have to dress it like you did for disk grinders to help them keep their shape.

Use dressing stones to help reshaping your wheels by grinding down any high spots and reshape the wheel profile. Use your tool rest to help keep the dressing tool steady and under control. Feed it into the wheel slowly to prevent gouging the wheel.

Wooden wheels are a little different from other wheels in that they are not made from a single piece of material. More commonly they are made of many pieces of cork that have been glued together to make the wheel. In fact this method of construction results in a wheel that is stronger than if it were a single piece of wood. Even though the glue is waterproof, such a wheel is used with only a slight slurry of polishing compound to do a rough polish of a piece. It needs to be treated carefully to maintain its correct shape.

Like almost all grinding equipment, spindle grinders will have a coolant or slurry supply system. This will feed it onto the wheel. Most of the systems will just use water for this. This is true for the diamond embedded and the stone wheels. The steel wheels will often use a grit slurry like used on the disk grinders.

Before starting to work, turn on the motor and then the lubricant to the wheel. Don't work it dry. Let the wheel get wet before you touch the piece to it. Then when you are done reverse the process. Turn off the lubricant to the wheel. Let the wheel rotate for a while (ten minutes or so) to let it spin dry. If you don't, the water will settle to the bottom of the wheel making it heavier there.

If you turn the grinder on at this point, the wheel will be out of balance and will vibrate back and forth as it rotates. This pounding will cause the bearings on the arbor to wear faster. It will also cause the wheel to surge against the casting and will make grinding very difficult. In the extreme, a highly unbalanced wheel may even break while you are working on it.

A wheel will throw some of the some of the coolant as it spins. Most of this spray is caught in metal hoods behind the wheel. Do not rely, on these to be safety guards. They will not catch everything. Coolant will inevitably get sprayed on you, so be sure to wear protective glasses and a studio apron.

Finishing Techniques 145

Figure 139. Using a small Dremel ® moto-tool.

It also helps to keep things out of your eyes if you keep the wheels at least a foot or so below your eye level. In order to keep as much material out of the air as possible, it helps to have your ventilation pulling air into the hoods and out of your shop.

Hand-held grinders

For surfaces that are not quite so flat, you may be better off using a hand-held grinder. The kind of hand-held grinders that most of us are probably familiar with are Dremel ® moto-tools like the one seen in Figure 139. Typically the ones you see in the store are single-speed models that run at 28,000 rpm or variable-speed ones that can be adjusted from 5,000 to 30,000 rpm. They will accept tools with shaft sizes up to $1/8^{th}$ of an inch in diameter.

You want to get a moto-tool, even if you are only drilling holes in your castings. Hand-held drills only rotate at about 1300 rpm. Diamond bits are made to be run at the 28,000 rpm typical of a moto-tool. At slower speeds, you are likely to get frustrated with the slow cutting speed and push harder on the tool

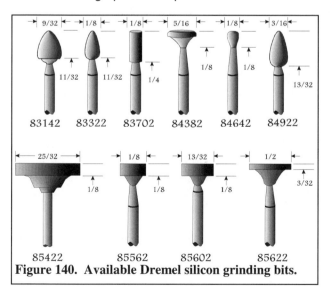

Figure 140. Available Dremel silicon grinding bits.

and ruin it.

Dremel also makes a heavy-duty model that will accept tools with diameters up to ¼ of an inch in diameter. This one, because it has a higher torque motor, can adjust its speeds down lower. It can be varied any where from about 0 to 20,000 rpms.

When using these tools, it helps to have a flex-shaft attachment and a variable speed control—a foot pedal type is the most convenient to use. This allows you better control and flexibility with the tool. Another benefit of the flexible shaft is that it gets the electric motor out of your hands. This is important because you should still work with these tools in a small dish of water to cool the glass.

For bits, you can use the gray-green silicon carbide grinding bits made by Dremel ®. They come in a number of different sizes as illustrated in Figure 140 and are good for smoothing but not much else. They tend to be too soft.

You can purchase better carbide tools and diamond-coated bits from sources such as Lapcraft Co., Crystalite Corporation, or Truebite. Truebite specializes in the small bits and we found them to be quite helpful. Either of these kinds of bits will have to be used wet and since a lot of silica loaded mist will be generated, we recommend that you wear a respirator with a combination mist/dust filter.

Using small hand-held grinders can be a little different than you might expect. Most people will push the tool against the glass and hold it there. This can cause excessive heating of the tool and cause it to wear. Lift the tool frequently to cool it and allow lubricant to release trapped glass from the wheel. Don't rush the job. Take you time. Let the tool do the work.

Grip the hand-held grinder tightly. Rest this hand against the piece or some other object to steady it. The helps you to control the position of the tool so that it does not run away from you and skip across the piece. It helps to have this hand raised above the level of the piece if possible. Always use a tool bit smaller than the size of the feature you are grinding for better control.

There are a number of different types of diamond grinding bits that you can get for hand-held grinders. They include small wheels, engraving bits, cylinders, drills, and points. With diamond ball tool bits, work

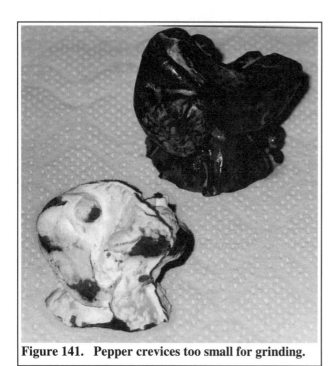

Figure 141. Pepper crevices too small for grinding.

they overheat. This practice will cause excessive wear even on plain steel wheels. The casting will be hurt because it will heat up excessively and may possibly crack. Lastly, your health will suffer because this will result in release of large amounts of glass dust into your breathing zone. Instead always ensure that there is a good film of lubricant flowing on the face of the cutting surface.

One of the first secrets to using grinding equipment is to hold your casting tight but not to push it into the tool too hard. If you do push it into the wheel too hard, the wheel can grab the piece and throw it across the room. Instead apply gentle pressure and let the piece ride on the cutting surface. This practice will also provide better lubrication to the piece reducing wear of the wheel. Move around on the wheel to prevent overheating one section and to even out wear on the wheel.

When grinding, go slow and just take off a little bit at a time. Do not let the glass overheat. Periodically dip the piece in water. Besides cooling it off, this will also clean it off so that you can see what you are doing. Cleaning off the casting in water is especially important when moving from a larger grit piece of equipment to a finer one. You do not want to carry over a bit of larger grit because it can contaminate the other piece of equipment so that you will get persistent scratches that are impossible to remove.

When grinding, any sharp edges are vulnerable to chipping. To avoid this problem, it is recommended to round or chamfer the edges on your piece just slightly to reduce this tendency. For those of you not familiar with the term, chamfering means to cut a little triangle out of the edge of a piece as shown in Figure 142.

the front of the bit more that the sides. Use reaming tools if you want to use the side of the tool.

Because of the finite size of the heads on the bits, you may find that there may be small crevices in a casting that you will not be able to get the grinding bits down into. Figure 141 shows both before and after shots of such a piece. It is a casting of a very curly pepper found in Jim's garden. Although much improved after grinding, it still has material down in crevices that were too small to reach into with the bits he had available. In a case like this, you may want to resort to sandblasting the last of the investment off the casting. If you have a sandblaster, it is probably easier to start using it from the start.

Besides grinding bits, you can also get many other smoothing tools for hand-held grinders. These include wet-dry sandpaper disks and rubber, cork or felt wheels. Any of these can be used with or without adding a slurry of abrasive or polishing compound.

Using grinding equipment

Never use a piece of grinding equipment dry. This is bad on the equipment, the casting, and your health. The equipment needs the lubrication to prevent damage. If it uses diamonds, they can melt off if

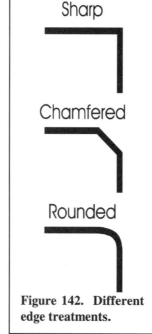

Figure 142. Different edge treatments.

The size of the chamfer does not have to be very big. A thirty-secondth of an inch is often effective without being distracting from the looks of the edge if it is going to be normally visible. You may want to make it a little larger on the edge of a base, which is not really visible and is more likely to receive rough handling.

After you are done grinding for the day, you usually want to dry off the grinding equipment. This is done to prevent any rusting or pitting of the

Finishing Techniques 147

wheels. To do this, you just turn off the lubrication to the equipment or empty the reservoir and let the equipment run for a while. Centrifugal force will force most of the lubricant off of the wheel. This is usually all that is necessary.

Sanding a casting

Another option for removing material from a casting is to sand it. The "sandpaper" that you will use is wet or dry silicon carbide or diamond bonded to a high-tech paper. Like grinding wheels it is available in a wide variety of grit sizes—anywhere from 60 to 1200. Again the grit size that you choose will depend on where you are in the finishing cycle. Start with large grits (low numbers) and move on to smaller ones. Remembering to clean the piece when going from one step to the next.

Sanding equipment

Like the other equipment we have discussed in this chapter, there are a number of different kinds of sanding equipment. They include belt sanders, drum sanders, disk sanders, orbital sanders, and hand sanding equipment. "Sandpaper" is readily available for each of these types of equipment. You can also get "paper" of other materials for this equipment to polish your work. These include cork, leather, canvas, and felt. So let discuss some of this equipment.

Belt sanders

When we say that you can use belt sanders on glass, we do not mean the kind of belt sander that you use on wood. The reason is that you have to work the piece wet. Wood belt sanders are not appropriate for this. A glass belt sander is built on an upright stand to which are mounted two rollers between which the belts are strung. The belt is supported from behind when you push on it by a metal back plate. A water feed system wets the belt from above. A catch and drain pan will be mounted at the base. The motor that powers the rollers is mounted behind the stand out of the path of the water.

Belt sanders come in many different sizes both in length and width. The length of a belt refers to the total length of the belt if it was cut open and laid out. Thus a 24" belt really only has a useful working surface of about seven inches. Belts are available in lengths from 18 to 48". They come in widths of 2½ to 5½ inches.

Belt sanders can be a little tricky to use. First, to try and sand a surface flat, you have to make sure that the back plate is riding right under the belt. If not, when you push the piece up against the belt and back plate, the belt will be raised at the top and the bottom edge. This will cause it to sand a little more off at the top and bottom giving the surface of the piece a slight convex shape to it.

The other thing that can be tricky when using a belt sander is to keep the belt centered on the rollers. The belts will tend to want to ride to one side or the other based on differences in tension of the belt from one side to the other. This is adjusted by using a screw on the top barrel that changes its tilt slightly. Sometimes you may have the belt riding perfectly centered and as soon as you touch your piece to the belt it starts to move on you. This can also be due to the back plate allowing too much play in the belt that then causes variations in its tension.

Disk sanders

Disk sanders are similar to disk grinders except that the sandpaper you mount to the disk does the material removal and not the disk. Disk sanders are available in size from about two to ten inches in diameter. They can be in the form of fairly hard fixed disks like on a grinder or soft rubber ones. The former will be a part of a table like system like for a disk grinder while the latter is more likely to be used in a hand-held mode.

The advantage of a fixed disk system over a normal disk grinder is the ability to be able to quickly change the grit size that you are working with. The backing disk will often be slightly padded to allow some give to it. This added play smoothes out the pressure on the disk and makes it a little less likely to pull the piece out of your hand.

The sandpaper disks are usually stuck to the fixed disk using some sort of adhesive. It allows you to attach and remove sandpaper disk a

Figure 143. Glass belt sander.

number of times. Normally though, it has been our experience that the ability to stick the sandpaper to the disk is the life limiting factor. They never seem to get worn out before we have to trash them. Denver Glass Machinery has some nice Velcro-backed diamond sanding disks by 3M that are good. They also have Roloc kits for Dremel moto-tools with 1" or 1½" sanding disks of 60, 120, 200, 400, 800 and 1800 mesh diamond grit.

You can also get flexible diamond grinder disks, which are permanently bound to a flexible disk. These are used in the glazing industry. The best tools to use these with are air grinders. These are like air drills but they also have a hookup for water that feeds water to the center of the disk as you are working on a piece. The reason that these are air powered should be obvious.

Drum sanders

These sanders are very similar to spindle grinders except that they use sandpaper belts on their rim to do the cutting. They are limited over spindle grinders in that they only come with flat ends. They are usually about eight inches in diameter and use belts or strips of sandpaper, which depends upon whether the drum is a vise lock or expandable type.

A vise lock drum has a split in the rim. The paper is run around the rim and inserted into the split. The split is then vise locked closed by flipping an adjusting lever. This lever tightens a bolt across the split that pulls it closed. This locks the sandpaper into place on the drum. You have to make sure that the sandpaper is pulled tight against the rim as you install it. If not, you may get a slightly raised section at the fold as it feeds into the split. This will cause this area to wear more and shorten the life of the sandpaper.

The other kind of drum, an expandable drum, uses small sanding belts. As seen in Figure 144, they are constructed of a heavy rubber rim mounted onto a metal drum. The rubber rim has grooves cut into it that allow it to expand radially outward just slightly as it is spun up. It is sized such that the belts will fit loosely when the drum is stationary but tightly when the drum is spinning. These drums should not be spun up without a belt in place. This will allow the rim to over expand and possibly tear itself apart.

Figure 144. Expandable sanding drum.

These drums are really convenient and allow quick changes of sandpaper from one grit to the next. They are available in six and eight inch diameters with widths from one and a half to three inches.

Orbital sanders

Orbital sanders are commonly used in the glazing industry to bevel or miter the edges of a piece of glass. They use diamond or silicon carbide pads that are either glued or attached with Velcro to the orbital plate of the sander. They do not remove material as fast as belt or disk sanders, but they are lighter and can be easier to handle. Orbital sanders are not meant for using to remove lots of glass.

Lubrication is usually supplied by hand misting diamond coolant with a spray bottle that you stop and use occasionally during the process. The pad sizes that we have seen are 3½" by 7¼" and 4¼" by 8¼".

These systems tend to be electrical ones. Since we use them wet, they should be used with a ground fault interrupt (GFI) circuit. This is to protect you against receiving any electrical shocks. If your manufacturer does not supply a GDI circuit on your equipment, you can make one using components from a hardware store.

Orbital sander use is relatively simple. After spraying on the lubricant, start with relatively light pressure to remove any sharp edges. Then gradually increase the pressure as the edges are

Figure 145. Using an orbital sander on a piece.

worn down. You can apply heavy pressure after all the sharp edges are broken down as long as you keep the piece well lubricated.

Hand sanding materials

Pads and sandpaper are also available for hand use. Wet and dry silicon carbide sandpaper is available from most hardware stores, but if you want the extra cutting ability of diamonds, you will have to order them special. The paper used for diamond pads is not really paper but a man-made fiber mesh.

Diamond pads are available where the diamond paper is bonded to sponge rubber pads. This makes them easy to grasp and conform to the surface that you are working. The sponge will also help hold and supply some lubricant to the surface as you are working it. Additional lubricant can be sprayed on or you can work on the piece in a dishpan of water.

Figure 146. Using diamond hand sanding pads.

You can also get "diamond files" as in Figure 147 where pieces of the diamond paper have been bonded to plastic handles or diamond embedded metal ones with different shapes. These allow you to apply a little leverage to the piece.

Hand working glass with paper, pads, and files is slow and tedious, but sometimes you can not reach into fine detailed areas any other way or you may have fine details that you want to retain. Take your time and try not to get too bored.

Using sanders on glass

As with any working of glass, always use lubricant when sanding on glass. We do not want to get glass dust into the air. Lubricant will keep the dust down and increase the life of the sandpaper.

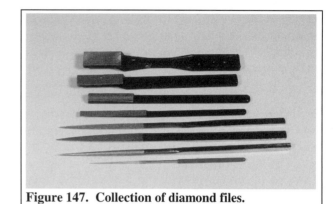
Figure 147. Collection of diamond files.

Because sanding paper, belts, and disks are not as robust as grinding disks, you need to treat them a little more gently. Of course they are also cheaper, but not so cheap that we can afford to waste them by abusing them. So always start off slow when working on any sharp edges so that they will not rip into the sandpaper. Once you have the sharp edges broken down somewhat you can increase the pressure on the glass. It also helps to get the power tools moving before bringing them into contact with the glass.

Always clean the glass when changing from coarser grit to finer grit. We do not want to get any coarser grit particles embedded into our finer grit paper. This will result in putting in new scratches as you are trying to get old ones out.

When using very fine sand or diamond paper with grits smaller than 800, always use light pressure. This will prevent breaking the small grit particles free from the paper and cutting deeper than you intended.

Buffing and brushing of castings

Buffing refers to an operation done near the end of the polishing cycle. It will use cloth and felt wheels. Brushing, on the other hand, can occur early in the finishing cycle to remove investment or later just before buffing.

Buffing and brushing equipment

The final stages of polishing a casting always consist of brushing and/or buffing. They both will give your work a higher degree luster than can be achieved by sanding or using a wooden wheel. Of course you only want to do this if you are trying to achieve a high luster polish.

© 2000 James Kervin and Dan Fenton

Brush wheels

Brushing with a brush wheel is often the last stage in polishing a casting. It will leave a semi-polished surface that many glass artists are satisfied with and do not go any further. Brush wheels are mounted to spindles like those used for a spindle grinder. The most common wheels consist of stiff bristles set in rows radiating from a wooden hub. They should be used with catcher hoods like other wheels.

Because of the flexibility of the bristles on a brush wheel they can often reach surfaces that grinders, sanders, and other wheels can not. This will allow them to at least put some polish on the nooks and crannies that you can not reach any other way except hand sanding. Depending on which polish you use and how hard you push they can provide a variety of final textures.

The brush is used with a slurry of water and a polishing compound like a fine pumice to provide the abrasive polishing action. Pumice is a rough, porous, gray rock that is formed in volcanoes. It is crushed and sold as a powder to be used for grinding.

Buffing wheels

To obtain an even higher degree of polish than can be obtained with a brush wheel, you can finish with a buffing wheel. The kind of wheel that we usually use is a muslin wheel. This type is made from a number of layers of unbleached cotton fabric or muslin that is stitched together either in spirals or multiple circles to hold the wheel together. Ones with about 50 to 60 layers seem to be the most popular.

The buffing wheels are coated with either a slurry of water and polishing powder or a wax/polish paste to provide the abrasive action. In either case, the wheel is dampened first. Slurries are mixed to the consistency of whipping cream or just slightly thicker and applied by brush to the wheel. Using a wax/polish paste is more conservative on polishing compound and allows it to stick to the wheel better. Do not apply the polish to the wheel too thickly or it will load up the wheel and actually inhibit polishing.

We suggest using one of two polishing compounds: tin oxide or cerium oxide. Tin oxide is a fine white powder that works well on harder wheels like felt, leather, or cork. Cerium oxide is a pinkish powder that works well on the intermediate to soft wheels of leather, felt, and muslin.

Because buffing wheels are soft and fluffy, people seem to treat their use a little more laxidasical. Don't, they can pull stuff out of your hands as easily as any spindle wheel. Be aware that they can also wear down your fingernails if you give them a chance.

Although the muslin wheels will not reach as far into the nooks and crannies of a casting as a brush wheel will, they still have a lot more penetrating power than a solid wheel or sanding belt. Buffing wheels come in a variety of sizes. With the motor speed being constant, the larger the diameter of the wheel the faster the edge speed of the wheel will be and the faster the buff will polish.

Before their first use, a muslin buffing wheel should be raked to roughen it up and remove any loosely bound lint. This would come off anyway in the first few minutes of work but since it can grab and tear at your work, it is better to get rid of it before you start. Make a rake by driving a half-dozen nails through the last couple inches of a comfortable length of wood. A piece about a foot or so is easy to handle for this purpose.

With the buff spinning, push the rake up against the buffing wheel. Hold it there for a couple of minutes. The rake will pull both lint and threads from the wheel and some will go flying. There is also the chance that if you push the rake too far into the wheel or do not hold it tight enough that it can get pulled out of your hands. For these reasons, it is suggested that you wear safety glasses and a dust mask when you do this. When you are done, stop the wheel and trim any stray threads with a scissors.

Figure 148. Raking a buffing wheel.

Finishing Techniques 151

Another common type of buffing wheel is a felt wheel. These are built up similar to muslin wheels in size and shape but they are much denser. Thus they can not reach into the nooks and crannies like a muslin wheel can. They will only hit the highlights. They do not need to be raked like muslin wheels. But after a lot of use, the outer rim can develop a cake of material that will have to be removed. Holding a heavy scraper perpendicularly up against the wheel can scrape off this coating. This will cause material to go flying so be sure to protect yourself.

The mounts for buffing wheels on some spindles are called threaded tapered spindles. They are tapered to a point and have either a right-handed or a left-handed thread on them. The left-handed thread goes on the left side of the spindle. You have to be sure to have the correct one on each side or the buffing wheel will loosen up and fly off the spindle as soon a you bring it in contact with your work.

Figure 149. Buffing wheel mounted on tapered spindle mount.

When using buffing wheels, it helps to think of them as sandpaper that you have to continually replace the grit on it. It rubs off as you are polishing a piece and you have to replenish it as needed. To do this just hold the buffing compound up against the wheel as it spins. With the wheel turning as fast as it is, it will only take a couple of seconds to replenish the wheel.

Hand brushing

Many times the first thing that you do after the casting has been removed from the mold and cooled off is to brush it down by hand. You are trying to remove all the investment material from the casting. How hard this is, is a function of what type of glass you used, how activated your investment becomes at the process temperature and how hot you actually went in the firing. Generally speaking, the higher temperature you go to in a casting firing the more likely you are to get investment sticking to your casting.

The easiest situation is where you have had virtually no mold-glass interaction and all you have to do is clean the dust from the mold off of the casting. For this situation, all that you need after breaking away the mold and picking off the small pieces with dental picks is to brush it down with a stiff vegetable brush. There are other brushes that you could use but these seem about right for general use. For those tight spots we also find an old toothbrush to be invaluable. Whatever you do, do not use a brass brush on a casting. The brass will rub off and stain the casting.

If you are the impatient type and want to speed up the brushing process, you may want to consider getting a brushing wheel. Such wheels, as explained earlier, are usually used for final buffing but that should not stop you from trying to use them also for initial cleanup. You may not want to use the same wheel for both purposes because you will lose buffing compound and you may get little chips off in an initial cleaning that could scratch a piece in final buffing.

Cleaning and etching compounds

When cleaning up a casting, you will sometimes want to use material to speed up the process. This can be an abrasive to wear away material or a chemical to help dissolve off the investment. Both of these can be useful additions to your cleaning bag of tricks.

Abrasive cleaners

The ideal abrasive for use on glass is Bon Ami. Bon Ami is a low-caustic, mildly abrasive scouring powder available in most grocery stores. Stronger abrasives like Ajax or Comet can scratch the glass. This is not the case with Bon Ami.

To use it, you just get the glass casting wet, sprinkle on a little Bon Ami and apply a little elbow grease with your scrub brush. (Please don't ask where to get the elbow grease.) After a good scrubbing, the casting should be clean with minimal dulling of the surface.

© 2000 James Kervin and Dan Fenton

Etching compounds

Okay, let's suppose that everything did not go perfect and you have a fair amount of mold material that interacted with your glass casting. The easiest way to try and remove it, although not always the most successful and definitely not the least hazardous, is to try a little better living through chemistry.

Many of the famous artists of old would use hydrofluoric acid to clean up their castings but we tend to avoid that stuff like the plague because of the great danger it presents if not handled carefully enough. Lime-Away will work to some degree to remove investment. It won't bring the casting up to a full shine but it may help remove some of the mold material.

Muriatic acid can also be used for this task with soda-lime glasses with some success but should be avoided like the plague with lead glasses, where it will dissolve the lead out of the glass leaving a soft porous casting that falls apart. This also points out why lead glasses should not be used for food bearing surfaces, especially acidic ones.

Muriatic acid is a diluted form of hydrochloric acid and will dissolve or corrode metal, so store it in plastic or glass containers. It is a strong acid and should be treated with respect. See some of the precautions listed in the back of this book for dealing with acids and read the instructions on the bottle. Use it to try and dissolve some of the investment off the surface by dipping the casting into the acid for a few minutes and then rinsing the acid off.

Although we have not tried it, we have been told that a solution of sodium thisulfate in glycerin also works well for removing investment residues. Both materials are also listed along with ammonium salts as good solvents for calcium sulfate (gypsum plaster) in our chemistry handbook.

Sandblasting

Another technique that you can use to clean up a casting is to sandblast it. Here you want to proceed very carefully because you do not want to lose surface detail. Use light passes over the casting to just lightly etch it. Make sure that you do the blasting in an enclosed cabinet. Otherwise you will be scattering silica dust all over the place. Not a good idea knowing what we do about breathing free silica.

If you are trying to only blast small details into a casting, then you may be interested in one of the small micro-sandblasters that are on the market. Rio Grande has a unit that is good for this purpose.

If you only want to do sections of a casting then you may have to mask off the sections that you do not want to etch. With flat glass, it is relatively easy to mask off a section of glass using clear self-adhesive shelf paper. With undulating casting surfaces, this is not so easy. If you try to apply sheets of this material, it will bubble and wrinkle. This can especially cause trouble if you then try to cut some of it away to expose a pattern.

Tapes work a little better than sheet material because you can form it to the surface of the casting better. Plastic tape like electrical tape will resist the blast well but it does not stick too well. For a light blast, masking tape works pretty well. It sticks well and is easily conformed to convoluted surfaces.

Even tape can be hard to completely conform to a surface and you will have the overlap areas to deal with. These can allow the blast to get in under the tape in these areas. So some times what you want are materials that you can paint on.

For this, glues like rubber cement or Elmers Glue™ are a good choice. We find Elmers the easier to use of the two. Rubber cement is hard to trim and it wants to stick to everything. You can make it a little less sticky by dusting it with talc or whiting after application. These glues are brushed on and then allowed to dry. They can then be trimmed up using an Exacto-knife. You can apply multiple layers for greater resistance to blasting.

Non-abrasive shining methods

If you don't like the matte appearance that you have on a raw or abrasively cleaned casting and you don't want to go through all the work of polishing it, you can make your piece appear more shiny and translucent by applying an oil or wax to it. These will give a sheen to the surface and also prevent any accumulation of fingerprints later on.

Oils and waxes

Oils and waxes will permanently change the surface texture of the glass casting making it look wet or damp. We have used a number of different oils for this purpose including salad oils. One spray oil that has been found to perform well for this purpose is

Varathane Natural Oil Finish #66 Clear. It is a clear oil that does not yellow after drying.

To apply an oil make sure that the casting is as clean as you can get it — clean from fingerprints, etc. Spray a good coat of oil over the entire surface being sure to get any deep crevices. Make sure that the entire surface is covered. Blot up any excess oil until the entire surface has a matte surface finish. If you see puddles in deep areas use the edge of a paper towel to blot them. If you get any dust on the surface wait until the oil is dry before trying to remove it. It will dry in about a half a day.

For a little harder and longer lasting surface finish you can use a wax. Almost any kind of wax will work — car waxes, furniture waxes and floor waxes. One that we find easy to use is Future floor wax. Apply it like an oil. Then after it is dry, buff it with a soft cloth.

Paints and varnishes

An alternative way to shine the surface of a kiln cast piece is to seal its surface with a clear acrylic or enamel. This will give the glass surface a wet look that is permanent. Like oils, the most convenient way to apply them is in a spray form. You might want to test these materials first on a piece of clear glass because some have a yellow tint to them.

To apply them, make sure that the surface of the casting is clean and free of oils before starting. You might want to wash it with soap and water and then air-dry it. Next, get your spray can of paint ready and apply the spray with thin even clean strokes. Start each stroke off the piece, go across the piece evenly, and finish the stroke again off the piece. Allow each coat to dry before applying the next until you are satisfied with its appearance. Do not apply too much at once or allow it to puddle in any of the crevices. This will allow it to run.

Overglazes

For soda-lime glass pieces, there is another alternative available to give your casting a permanent shiny surface. You can try using a low firing-temperature clear glass enamel, often referred to as a glass flux, to shine up the surface. For this we recommend Flux 92 by Standard Ceramics Supply, Inc. or Back Magic from Fusion Headquarters. So named because it is used to make the backs of fused projects smooth and shiny.

Either will mature with a 15 to 30 minute soak at 1100°F.

This technique will not work for lead-based glasses because the casting will slump at these temperatures. Be sure to ramp the kiln temperature up and down very slowly if you have much variation at all in thickness of your piece so that you do not thermal shock it. Jim has had this problem on more than one occasion and nothing is more frustrating than cracking a piece in the final finishing step. Since these low firing-temperature fluxes are high in lead content, you will not be able to use them on anything that might be used as a food-bearing surface and there are no comparable non-lead bearing products.

Some artists don't like to use overglazes on their work because they say that these low melting temperature fluxes leave a hazy finish on their work. Instead they may "fire polish" their work. To do this they heat up the work in a kiln just to the point where the surface of the glass starts to get soft and smoothes out. This can be a real tricky process though because you get really close to the slumping temperature of the glass.

If you are going to try this technique, you will need a very accurate and responsive kiln controller to avoid slumping problems. We have been advised that this can be achieved with Bullseye glass cast pieces by heating them into the range of 1275 to 1285°F for about one half hour. We suggest careful calibration of your controller to assure the proper temperature control prior to trying this. Be aware that large pieces may slump when you do this.

Also you may not be able to flux or fire polish all the surfaces of your casting because at least one surface has to be on the kiln shelf. That is of course

Figure 150. Cast beads being fluxed on a bead stand.

unless you are making small items like beads, which can be hung on a ceramic bead stand as shown in Figure 150. If you think about it, you probably also realize that you could decorate your work with low firing-temperature enamels which would allow you to add decorations to the exterior of your piece instead of fluxes. This idea may shock many pâte de verre purists but what the heck. You are the artist—you decide.

Safety

Safety should always be your number one priority during kiln casting even above producing a quality product. The best way to stay out of trouble is to understand your materials and equipment. This understanding will allow you to realize where the potential dangers lie in your work and how to minimize them. In this chapter, we will remind you of many of the hazards that we have touched upon previously and provide you with some general safe operating rules.

Working around rotating equipment

As discussed, the concern here is that you will get yourself caught up in the equipment. Do not wear excessively loose clothing. Tie back your hair. Do not have anything hanging down in front of you. Ties from aprons should be behind you.

Another danger with this kind of equipment is catching of hands and fingers. These can easily get caught between belts and pulleys. This can result in strains, cuts or even dismemberment. Jim's dad almost lost a finger that way once, so don't let it happen to you. Be aware of where you place you hands. These spinning objects are also to children like lights are to moths. They are irresistibly drawn to them. It is thus recommended that all belts and pulleys have guards over them that can prevent any access of hands and fingers to them.

Cuts

Castings fresh from the mold will invariably have sharp edges that can slice you real easily. So it is recommended that you wear light garden gloves while you demold and during the initial cold working steps. Be careful because looks can be real deceiving.

In kiln casting, you will also be working with a number of very sharp tools. These include modeling tools, knives for mold frame construction, and casting cleaning tools. Any of these tools will get you when you aren't paying close enough attention.

Treatment for cuts

Most of us know how to treat cuts. You stop the bleeding, clean the wound, and bandage them up. Sometimes you will apply antibiotics to prevent infection or get a tetanus shot if it was a puncture wound. For large cuts, go to the emergency room and get stitches.

Fires

In working with your kiln or hot plate, a number of general rules should be followed to avoid starting fires. First is the proper maintenance and location of your equipment. So check it out occasionally to see that everything is working correctly. Let's now reexamine some of the general safety rules for setting up your kilns.

Positioning your equipment to avoid fires

1) Mount a kiln on a stand to allow air circulation around it. Position your kiln at least one foot away from sheet rock walls and two feet from exposed wood.

2) Put a non-flammable work surface underneath your kiln. Concrete is best, ceramic tile or brick

is second best and some sort of non-asbestos fire-resistant board is the minimum requirement.

3) Make sure that no flammable liquids are stored near your kiln. If some are stored in the same room, provide sufficient ventilation to prevent build up of vapors and store them in a flammable liquids' storage cabinet. It helps to have floor vents to carry away the heavy vapors from such materials.

4) Check all natural gas sources in the area, if any, to make sure that you have no leaks. Do this by painting all the joints with a soapy water solution.

5) If you work in your garage, back your car out before you start working so that there are no gasoline fumes around.

6) Try not to leave your kiln unattended. If you must leave your kiln, try to pop in and out to check on it. If you are the forgetful type, we suggest using an interval timer to remind you to check up on it. Even with kilns on a controller, problems have been known to happen. Dan had his controller go crazy one night, possibly because of a power surge and returned to find the kiln glowing bright.

7) Know where to go and how to remotely cut off the power to your kiln. This will most likely be at your fuse or breaker box.

8) Always keep at least one ABC rated fire extinguisher nearby. Position it so you will not have to reach over a fire to get to it, preferably near an exit, and know how to use it.

9) Keep a clear exit from your work space at all times through which to escape in case of a fire.

10) Have your kiln properly set up and fused or breakered by a qualified electrician. It is best to have a separate dedicated circuit for each unit. If you have more than one kiln on the same circuit, only run one at a time.

Proper operation of your equipment to avoid fires

1) Never allow children to play around your equipment.

2) Do not allow combustibles to build up in your work area.

3) Read and follow the manufacturer instructions for your equipment. Never lay anything on a kiln that you would not be afraid to stick into a kiln. The "cold" face of a kiln can get hot enough to ignite paper (remember 451°F for those science fiction fans out there) and other materials.

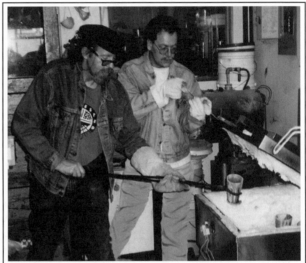

Figure 151. Have a door jockey when doing crucible work.

4) When working with crucibles, never work alone. If nothing else, your assistant can act as door jockey as Jim is doing for Dan in Figure 151, and let's hope that's all. But that way if anything does go wrong, there will be someone there to assist.

5) Be careful in heating wax so as not to get it so hot that it can flash into flame. A double boiler arrangement is the preferred heating configuration. Remember not to let the water boil out. Do not use open flame burners for heating wax.

6) Never put a 30-amp fuse into a 20-amp receptacle or you're just asking for trouble. Also if you're not getting enough juice out of your kiln, do not switch your breaker for a higher service one. These are intended to be the weakest part of the circuit and have been sized so that the house wiring does not burn or catch fire.

Safety

What to do in case of a fire

Besides following all of the safety precautions listed above you need to understand how to extinguish a small fire so that it does not get a chance to grow. To do this you must, as was mentioned, have a good fire extinguisher. The right kind of extinguisher for our work is an ABC dry chemical extinguisher often referred to as a tri-class dry chemical fire extinguisher as seen in Figure 152.

This type of extinguisher uses nonconducting chemicals (good for use around kilns) that effectively suppress the three main types of fires. As you may know, these are:
(A) fires involving combustible materials like paper, wood, cardboard, cloth or other similar materials;
(B) fires involving oils, paints, gasoline, chemicals or other flammable liquids; and
(C) fires involving live electrical equipment, such as kilns, crock pots, etc.

ABC fire extinguishers are available in easy-to-store wall mounted units weighing between 2½ to 30 pounds.

You should read over the instructions that come with your fire extinguisher and ensure that you understand how to use it. General rules for operation and use of a typical ABC dry chemical fire extinguisher are as follows:

1) Hold the extinguisher upright during use so that the dry chemicals feed properly.

2) Most extinguishers have a safety pin that you have to remove first to allow you to pull the trigger and start dry chemical flow.

3) Don't get too close to the fire. Stand about eight feet away from it.

4) Aim the extinguisher at the base of the fire and squeeze the handle to start dry chemical spray.

5) Sweep the chemical spray from side to side to completely cover the fire.

6) If your fire gets too big, evacuate the building and the nearby area. Notify fire department.

Figure 152. Dry chemical fire extinguisher.

In order to ensure that your fire extinguisher will work when you need it, you should inspect it periodically. This should be at least monthly.

They will usually have a pressure gauge, which should read in the proper range.

If you ever use your extinguisher for any reason, have it recharged and checked out by an authorized distributor since they are more prone to leakage after once being used.

After use, clean off all surfaces coated with the dry chemical because it is fairly corrosive.

Burns

Burns can be caused by exposure to caustic chemicals, electricity, radiation or heat. In kiln casting of glass your main concern will be hot equipment and wax, although we may occasionally use chemicals or can be exposed to electricity. Burns as you may know are classified by their severity as first-, second- or third-degree burns as follows.

- **First-degree burns** are usually characterized by reddening of the skin and pain. This type of burn, although painful, will heal fairly quickly.
- **Second-degree burns** are characterized by the development of blisters and swelling. They heal a little slower but are usually not serious enough to require medical attention.
- **Third-degree burns** involve damage to deeper skin layers and may have a charred appearance. They are often not very painful because the nerve endings in the skin may have been damaged.

You should always seek medical attention after a third-degree burn because of the damage that has been done to the body's protective layer to infection.

What to do to avoid getting burns

In kiln casting, burns are usually a result of inappropriate attire or not paying proper attention to what you are doing. Here are some general purpose rules to help avoid situations that can result in burns.

1) When working around hot kilns protect your body parts from the heat by using gloves, long

© 2000 James Kervin and Dan Fenton

sleeve shirts, light jackets, long pants, and closed shoes. These clothes should be natural fibers like cotton or wool, not man-made ones like nylon which shrink when exposed to heat earning them the pet name of "shrink wrap."

2) For high temperature work use Kevlar or Kevlar PBI gloves. The "Zetex" gloves are a type of Kevlar. Stay away from asbestos at all costs. Some examples of the kinds of high temperature resistant glove that you should use are shown in Figure 153.

3) Always turn off a kiln before you reach into it. We seriously suggest the installation of a dead man's switch on the top of the kiln that automatically turns it off for you whenever you lift the lid.

4) Be aware of which objects are hot and where you may have set them down so that you will not touch these areas until they have cooled. Even the outsides of gloves can become hot enough to burn you after reaching into a kiln.

5) Be careful when heating wax. Be sure that it is dry when adding it to the wax pot. Water will sink to the bottom of the pot and flash boil causing wax to erupt from the pot. Put a cover on that pot.

6) Don't splash wax around. Better yet don't even carry it around when liquid, except in your injector. Be careful when injecting wax that it does not squirt out in unintended directions.

Whenever you are handling acids like muriatic acid or hydrofluoric acid to clean your casting, you need to wear special equipment to prevent receiving chemical burns. The following list of protective gear is suggested if you decide to use acids. Of course how much of this you wear depends on how much acid you are handling. We suggest considering splash goggles, gloves, and a rubber apron as the minimal protection necessary.

1) Splash goggles that seal to your face should be used to protect your eyes with a face shield over the goggles to protect the rest of your face.

2) Rubberized clothing, which may include: laboratory apron, shirt, hat, boots, and full length rubber overalls that fit down over your boots.

3) Gloves appropriate for the material being used.
- Natural rubber is good for dilute acids, alkalis and alcohols.
- Neoprene rubber is good for dilute acids, alkalis, alcohols and ketones.
- Butyl rubber is good for acids, alkalis, alcohols, ketones, esters and many other solvents.
- Nitrile are good for dilute acids, alkalis, petroleum solvents, oils, grease and amino acids.

You may also want the gloves long enough to roll the top down to make a cuff to prevent solvents from running down your arm if you are not wearing anything else rubberized under the gloves (such as a shirt.)

4) Of course, a respirator with appropriate mist or acid cartridges is mandatory.

Figure 153. Temperature protective gloves for kiln work.

What to do if you get a burn

Treatment for a burn depends on the type and severity of the burn. Very severe burns may also involve shock. This is a protective mechanism of the body to reduce the demands it places on its component systems. Factors that may contribute significantly in inducing shock include:
- loss of blood or chemical balance
- extreme pain
- traumatic experiences.

A person in shock may look pale, feel clammy and be nauseous, dizzy, disorientated or headachy. In the event of shock, you should lie the victim down to allow better blood circulation, cover them to preserve body heat, and administer fluids if

Safety 159

requested in the form of sips of water or water mixed with baking soda to the ratio of 1/2 teaspoon to a quart of water. Always seek medical attention immediately if shock is suspected.

Let's now look at what you can do in the way of administering first aid treatment for the different types of burns.

Thermal burns

First-degree burns are treated by cooling off the affected area by running cold water over it. Then if necessary apply a dry compress to protect the area.

Second-degree burns can be treated by immersing them under water for up to a couple hours. Then, if desired, use cold compresses to cover a second-degree burn. After you are done cooling them, blot them dry and cover them with a sterile compress. Do not apply antiseptic preparations, ointments, or sprays if the burn is severe. Also never pop blisters or intentionally remove skin from a burned area because this opens you up to infection. For large burns you may also want to elevate the affected limb to prevent swelling.

Third-degree burns are usually just stabilized and then medical attention is sought. Do not clean the burn or try to remove attached clothing. Cover the burned area with a clean compress of freshly laundered material and then possibly cover this with a clean plastic bag to keep out germs. Elevate the affected limb. Have the person sit down and not walk if possible. Be wary of the possibility of shock. Seek immediate medical attention.

Chemical burns

If the chemical burn is to the skin, quickly rinse away as much as possible of the chemical with large amounts of water using a shower or hose for at least 5 minutes. Remove any articles of clothing from the affected area. Do not scrub the area. If any directions for treatment of burns are present on the chemical container, follow them. Apply a bandage and get medical attention. Notify medical personnel what material you were working with or better yet bring the can or label off it with you.

If the burn is to the eyes, rinse them with large volumes of water as quickly as possible for at least 5 minutes for acid burns and 15 minutes for alkali burns. (Alkali burns are more tricky because the eye might appear at first to be only slightly injured but this can progress to develop deep inflammation and tissue damage.) While washing out the eye, ensure that the face and eyelids also get washed since they probably have been affected. If only one eye is affected position that eye down and rinse from the nose outward so that you are not transferring material into the other eye.

For acid burns, if a weak solution of 1 teaspoon of baking soda in 1 quart of water can be made in a timely manner, rinse with this solution after the water rinse. For alkali burns, check the eye for any loose particles of dry chemical and remove them with a sterile gauze or a clean handkerchief. Rinse some more and then cover the eye with a dry pad or protective dressing. Try not to have the victim rub their eye. Seek immediate medical attention.

Electrical burns

Electrical burns are treated like thermal burns. With the exception that shock and CPR treatment may be necessary. You should consider getting CPR training from the Red Cross for just such an occasion. Be sure to turn off any live electrical equipment before touching a victim if he or she is still in contact with it. Otherwise you may become a second victim.

Toxic materials

The first step in evaluating what type of material toxicity plan you need for your studio is to examine which materials you are or will be using in your work. What are they? What is in them? Are they hazardous? How hazardous? What form do they take? How might they enter my body? The more informed you are about these materials the better you will be able to evaluate what you need to do to protect yourself, your employees and your family.

As an example, investment materials tend to be respiratory irritants at the very least. Free silica exposure (silica, as you may remember, is used as a refractory additive in many investment formulations) if inhaled can lead to a disease of a progressive nature called silicosis after heavy exposures of only a few months duration. Of course we are talking of heavy industrial exposures here, but the same thing can happen from low level exposures, it just takes more of them over a longer period of time. Thus, it is important that you observe good hygiene practices whenever working with molds and investment materials.

If you mix any of your own colored glass, be aware that many of the colorants are toxic materials. They are usually heavy metals, like cadmium, which are poisonous and not easily eliminated by the body. Operations like scrapping kiln shelves or grinding glass also creates dust-containing silica.

Other bad actors include: fiber paper that has been fired to high temperatures, overglazes since they are frequently high lead glasses, enamels and paints since they contain lead and other heavy metals, very fine frits, plaster, cements, etc. Even things that seem as innocuous as vermiculite can pose a health hazard. Its chronic inhalation can cause asbestosis-like reactions or cancer and ingestion can also cause cancer.

There are a number of different designations for toxins based upon how this toxin attacks the body. They include:
- <u>Poisons</u> are toxins that interfere with chemical processes in the body. These toxins may be cumulative or non-cumulative depending on how they are eliminated from the body.
- <u>Mutagens</u> are toxins that cause changes in your genetic blueprints.
- <u>Teratogens</u> are toxins that affect how the genes in a developing fetus become expressed by changing the background chemistry of the environment.
- <u>Allergens</u> are toxins that react with the body's immune system. These reactions may vary from as minor as the sniffles and watery eyes to as major as anaphylactic shock.

The degree of toxicity of a material describes its capability to hurt you. Highly toxic materials may only need a little bit to cause problems. So you can not always judge danger by how much material you are using. You also have to consider how often you use the material. Long-term usage of less toxic materials might also lead to health consequences. As an example, cancer is now suspected to be a result of chronic exposures to low levels of many chemicals called carcinogens. In fact, it is not understood if there are any completely safe exposure limits to carcinogens.

So the best practice is to minimize possible exposures to all toxic materials. To accomplish this, you have to understand how toxic materials can get into your body. There are three main routes of entry into the body: skin contact, ingestion and inhalation. We have already discussed how to avoid skin contact with toxic materials when we discussed what to wear to avoid chemical burns — gloves, glasses, etc. Ingestion can be avoided by restricting eating, drinking, chewing gum, and smoking from the work place. Inhalation is harder to control because we can not restrict breathing. Let's look at what you can do to avoid problems with toxic materials in your work.

Good work practices to avoid toxicity problems

One of the best ways to avoid problems with material toxicity is to develop good work practices that incorporate a high degree of cleanliness. Examples of good work practices when dealing with powdered toxic materials like those present in investment mixes are as follows:

1) Isolate a work area for mixing these materials so the chemical powders do not get spread all around your studio. (If you do not have the space to dedicate to the mixing process, create one temporarily when mixing the materials and then clean it up right after you are done working with them.) This does not in any way mean that you should not keep an isolated work area clean. This work area should have adequate ventilation to help remove airborne dust without at the same time tending to create it.

2) Consider using even more restricted volumes for the handling of your more toxic materials to help contain them. You could mix colorant powders in a small home-made glovebox, as illustrated in Figure 154, or confine all spray painting to a paint hood to help minimize your exposure to them.

3) If you have two processes using the same toxic material, try to collocate them so the materials do not get spread all around the work place.

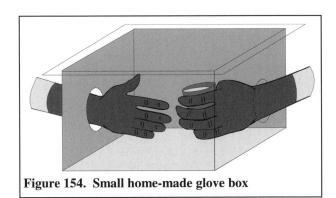

Figure 154. Small home-made glove box

4) Clean your work place regularly by either wet mopping or using a HEPA vacuum cleaner. Other methods such as sweeping or using an ordinary shop vacuum will just spread the dust around into your breathing air and contaminate your whole work space.

5) Store all materials in covered containers like those seen in Figure 155 to prevent them from being spread around. This also helps in preventing some of these materials, like plasters, from absorbing water from the environment.

6) Make sure that all the materials are properly labeled as to what they are and what hazards they present. Obtain and keep on file Material Safety Data Sheets (MSDSs) for all toxic materials that you use in your work. Understand how to read them and what they mean.

7) Whenever you have the choice between using either of two materials, use the least toxic one. Such as using non-asbestos rather than asbestos gloves in kiln work.

8) Have a training program for both you and your employees so that you are sure they understand the hazards of the materials that they are working with and proper work practices for their use. Besides protecting them, this will also help protect you from them.

9) When mixing investment materials, lead based glasses or colorants, you should always wear a respirator as Jim is doing in Figure 155 rated for the type of material and process you are engaged in. Make sure that your respirator fits and that you are wearing it properly. Respirators are not 100% effective and should not be substituted for proper ventilation.

10) Acquire and have on hand at all times, the proper safety clothing and equipment for handling the corrosive materials that we have discussed previously if you choose to use them. These may include the following: face shield, goggles, gloves, aprons, safety shoes, etc.

11) Do not eat, drink, chew gum or smoke in your studio. You shouldn't smoke anyway, but don't compound the problem by having that habit result in ingesting other unintended materials at the same time.

12) Don't take toxic materials home with you: leave them in your studio. Wash your hands as you leave the shop or as you arrive home. (Please, not in the kitchen sink.) Change your clothes before leaving the shop, or at the very least, wear a long shop coat over your normal clothes which you leave in the shop. Wash and dry work clothes separately from family clothes.

13) Dispose of all chemicals properly to avoid exposing your friends and neighbors.

14) If you have a cat, keep the critter away from the work area when dusty or toxic powders are being used. Cats lick themselves clean and will ingest all of the nasty stuff with which it comes into contact. Dan almost lost a cat this way from exposure to toxic colorants during one pâte de verre class. Don't worry as much about dogs because they only bite mail carriers and eat homework.

Ventilation

As we have discussed, the route for entry of toxic materials into the body that is probably hardest to guard against is inhalation. Whenever materials are heated to the melting point and above, mixtures of particulate matter and gases are released. The worst of these are probably metal fumes.

Figure 155. Always wear a respirator when mixing investment.

Inhaling large amounts of metal fumes can cause both immediate and long-term effects. Acute exposures can cause flu-like symptoms within about 6 hours that can last for up to 24 hours. Chronic low-level exposures to toxic metals such as lead can cause permanent neurological damage and reduced brain function. Small amounts of lead fumes are released during high temperature processes such as when using low melting-temperature lead glasses and fluxes.

Some of these fumes are heavier than air, so you should ventilate your workspace both high and low. You should consider direct ventilation of your kilns. Kiln ventilation systems that draw air out through a hole in the bottom of your kiln seem to be the most effective. Burning wax out of your molds releases toxic fumes such as acrolein and formaldehyde and it is better if they do not get out into your work space.

Before we discuss how to guard against inhalation problems, let's devote a little time to understand what we are guarding against. Inhalation hazards can come in a number of different forms: gases, vapors, fumes, mists and dusts. Do you know the differences between them? They are defined as:

- **Gases** are materials in the air that we breath that do not have a solid or liquid form at normal pressures and temperatures.
- **Vapors** are gaseous forms of solid or liquid substances such as steam that may or may not be the result of increased temperature.
- **Fumes** are extremely small solid particles created by heating metals to above their melting point or by chemical reactions.
- **Mists** are small droplets of liquid which are released into the air by mechanical actions such as spraying, splashing, or bubbling.
- **Dusts** are solid particles of many sizes and shapes that are generated by mechanical actions such as grinding, crushing, sanding.

The smaller or more gaseous a material, the easier it is for that material to penetrate deep into your lungs and be absorbed by your body.

Ventilation tries to solve the problems caused by these materials by one of two methods: diluting them or removing them.

- **Dilution ventilation** is where the objective is to try and reduce the concentration of toxic materials in the air we breath by mixing it with large volumes of fresh air. This is accomplished by exhausting large volumes of air and replacing it with new air. This does not mean installing a room air conditioner in a shop window. They do not exchange anywhere near the amount of air that we are talking about.

- **Local ventilation** in most cases is a better solution. Here an attempt is made to remove the toxic materials from the workplace at their source by sucking them out at the point where they are generated such as in ventilating the kiln.

When considering ventilation plans, here are some general principles to consider:

1) Try to remove contaminated air at its source. Pull it away from you and your employees. It is easier to control contaminated air by pulling it out of your workspace rather than by trying to push it out. Put a hood over your kiln.

2) Exhaust systems are more efficient if they minimize the distance they have to move the air. Have them exhaust the air outside of the shop as far from air intakes as possible and be sure to discharge your effluent air in a responsible manner. Do not pollute someone else's air.

3) Airflow is more controllable if you avoid unwanted cross drafts and add sufficient supply make-up air through a planned intake system.

4) Ventilation systems work best if they are under negative pressure. This is accomplished by positioning the fan at the exit of the system not at the source.

If you are really interested in installing your own ventilation system, we suggest that you contact an industrial ventilation specialist or get the reference book by Clark, Cutter, and McGrane on this subject listed in the reference section of this book.

Respirators

Sometimes when doing especially dirty operations or when doing temporary dirty operations in a general workspace, you may want more protection than is offered by your ventilation system. This is the time to use a respirator. It should be stressed that respirators are usually considered temporary measures and should be used as the primary protective device only when no other means is possible or the contaminant is so toxic that a single control measure is not felt safe enough. Your primary means of inhalation protection should always be your ventilation system.

If you are an employer and decide that your employees need to periodically use respirators, you should know that OSHA requires a written plan for their use. This plan is to acquaint your employees with respirator use and selection. It should include a medical screening to look for proper respiratory system function, a fit check of the respirator, training in their use as well as their limitations, and setting up procedures for their maintenance.

Choosing a respirator

In choosing a respirator, you need to first consider what form of airborne toxic material you are trying to protect yourself from. Does it contain dust, mists, fumes, vapors or gases? Next you should consider how long you may be working in that atmosphere. Longer times require larger respirator canisters. How toxic is the material? Lastly you need to look at how the respirator fits. They should be comfortable, leak proof, easy breathing, and non-interfering with vision. Some of these questions may be difficult for you to answer. If so, you should consult a reputable respirator consultant.

The type of respirator that most artists will use is a quarter mask respirator that covers only the mouth and nose as shown in Figure 156. The figure shows a front and inside view of a typical quarter mask respirator as well as a part break down of it. The individual components shown are:

A. Headband assembly
B. Face piece
C. Exhalation valve seat
D. Exhalation valve flap
E. Exhalation valve cover
F. Cartridge
G. Assembled cartridge adapter assembly
H. Gasket
I. Cartridge adapter
J. Inhalation valve flap

To get proper protection from your respirator, it is essential that you have the proper fit. Otherwise instead of the inhaled air passing through the filter, it will take a shortcut through a leak path.

To do a respirator fit, put some chemical filters in place, fan some strong smelling chemical vapors for which they are rated near your face or stand in a bagged off area saturated with the vapors and see if you can smell them. Be careful not to splash anything into your face. (You may want to wear some splash goggles when doing this to protect your eyes.)

To ensure that you have a good fit, you should exercise the seal. Do this by moving your head around—both side to side, and up and down. Frown. Smile. Talk. Bend over. Lean back. Breath heavy. Any of these operations could potentially cause the seal on your mask to leak.

You should also do a quick respirator fit and leak check every time you don the respirator. You first do an inhalation check by holding you hands over the ends of the filters and sucking in. You should feel the mask pull against your face. The longer the face piece remains pulled in the better the fit. It should remain for at least five to ten seconds.

Next do an exhalation check. Hold your hands over the exhaust valve and exhale. If there are no leaks, this should cause the mask to lift up off of your face.

If both of those tests are successful, your respirator fits and you have a good seal. You guys with beards and/or long side burns should be aware that there is no way that you are going to get a good seal and that the respirator will not be completely effective for you. The fit of your respirator should be checked periodically to ensure that physical changes such as weight loss or gain have not altered it.

Figure 156. Different components of a respirator.

Once you have a good fitting respirator, you need to decide what type of filters that you need. As should be obvious, this is a function of the materials with which you are working. Dusts and fumes usually just require mechanical filtration of the air. This is provided by a filter made of folded felt or paper. Mists, vapors and gases will require cartridges incorporating chemicals that purify the air by trapping the materials. Some examples for which chemical cartridges are available include: acid gases, organic vapors, paints and ammonia.

Respirator care

Like any piece of safety equipment, respirators require periodic maintenance to ensure their proper operation when you need them. After each use, they should be cleaned and checked for worn or broken parts. Look for deteriorating plastic surfaces, rusting metal, and cracked glass.

If okay they should be stored in plastic bags. The plastic bags ensure that chemical absorbents are not being exposed to atmospheric vapors when not in use. This can greatly reduce the useful lifetime of a respirator cartridge because they become saturated with these vapors even from air exposure without use. For the same reason, careful records should be maintained of the amount of time that they have been used.

Respirators should be stored in a place that is out of the way but at the same time is easily accessible in the event of an emergency. This location should also protect them from sunlight and temperature extremes. You should always keep replacement cartridges and filters on hand and they should be stored similarly.

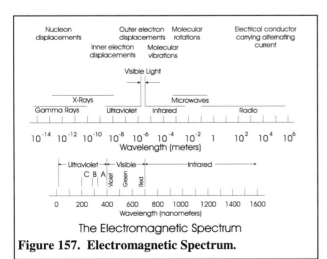

Figure 157. Electromagnetic Spectrum.

You can usually tell when filter cartridges are starting to reach the end of their useful life because they start to get harder to breath through as they get clogged up. Chemical filter life is not so easy to judge. They just stop absorbing the chemical agents without any warning. You might be able to test using less toxic aromatic chemical agents but sometimes not.

For this reason, they are usually considered to be consumed after 8 hours of use or two weeks after being unpacked. Even if never removed from the package, some chemical cartridges may become ineffective over time. Such cartridges will usually come stamped with an expiration date. Also respirators are designed to be effective for what are considered normal exposures and will just not be effective against very high concentrations of contaminants. They will not be able to absorb all the material and will wear out quickly.

Optical considerations

In working with high temperature kilns, beside fumes and toxic gases you also have to worry about more esoteric things like exposure to non-ionizing electromagnetic radiation. This radiation is in the light we see and that which we do not see. It totality it is known as the electromagnetic spectrum.

It is the result of electrons that have been thermally excited to higher energy states falling back down to lower energy states. Each transition from energy level to energy level results in the release of a photon with an energy equal to that of the difference between the two energy levels. This cycle is repeated as long as heat is applied to the glowing material.

Electromagnetic spectrum

The electromagnetic spectrum is just what its name implies, a complete spectrum of radiation with wavelengths which, as illustrated in Figure 157, range from angstroms to meters. From examining the upper chart in the figure, you can probably guess that the shorter the wavelength the more damaging the radiation. The area of the electromagnetic spectrum of concern for us is that from 200 to about 2000 nanometers A nanometer, abbreviated as nm, is 1/1,000,000,000th of a meter (10^{-9}). The lower cutoff is 200 nm because radiation with wavelengths shorter than this are usually effectively absorbed by air.

The different portions of this spectrum have been labeled based on some combination of properties such as wavelength, common use or biological activity.
- **Visible light** is that portion of the spectrum that humans have evolved to use in sight. This familiar span ranges in wavelength from about 400 nm for violet to about 700 nm for red light. The exact range visible to each individual varies slightly.
- **Infrared radiation (IR)** is that radiation with wavelengths longer than visible light. It is usually associated with heat energy.
- **Ultraviolet radiation (UV)** is that radiation with wavelengths shorter than the visible light.

The ultraviolet region is usually further subdivided into smaller bands on the basis of phenomenological effects. Because these effects do not have sharp wavelength cutoffs, the effects may carry over somewhat between bands.
- **UV-A** is the band from 320 to 400 nm. It is that spectral region that was used in the 60's to excite those fluorescent posters.
- **UV-B** is the band from 290 to 320 nm. It is that band of radiation usually known for causing sunburns.
- **UV-C** (200 to 290 nm) is the lowest wavelength band. It is not present in great quantity in nature because it is absorbed by earth's atmosphere.

Electromagnetic radiation damage

When a light photon is absorbed by tissue, all of its energy is transferred to the absorbing atom or molecule. This energy puts the atom or molecule into an excited state — the mode of which depends upon the wavelength of the photon. The shorter the wavelength, or as you may remember the higher the energy of the photon, the more energetic the result.

In increasing order of energetics the following may occur: molecular rotations, atomic vibrations in the molecule, changes in electronic energy state or expulsions of electrons. This highest energy result is known as ionization and usually takes more energy than is available from the region of the spectrum with which we are concerned. That is why we originally said we are dealing with non-ionizing radiation.

Infrared radiation may induce rotational and vibrational states in a molecule. Visible and ultraviolet radiation may induce higher vibrational states or electronic excitations. Vision and photosynthesis are two beneficial results of such interactions. Once the energy is absorbed by molecules within a cell any of a number of results are possible. The energy could be dissipated as heat. The molecule may be structurally altered or even break apart. The molecule may react with another molecule. The overall result to the cell from these changes may vary from changes in cell function to death. The question is what effect does this have on the eye.

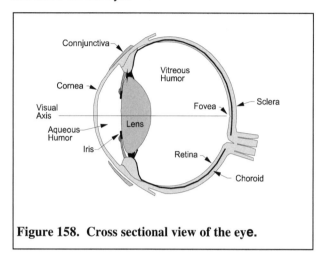

Figure 158. Cross sectional view of the eye.

The eye, as illustrated in Figure 158, is essentially a near-spherical organ with a transparent window on the front called the cornea. The cornea is part of the exterior sheath of the eye called the sclera. The eye is sheltered from the outside by the brow and the eyelids of which the conjunctiva is part. The iris divides the eye into two chambers filled with clear fluids or humors. It also serves to regulate how much light enters the eye.

Rays of light enter the eye and are focused first by the shape of the cornea and second by that of the lens onto the retina. This is where the chemical reactions that the brain interprets as sight occur. Electromagnetic radiation of different wavelengths interact differently with the eye. Let's examine how these different spectrum ranges can damage the eye.

UV-C is essentially totally absorbed in the cornea. Its effects wear off relatively quickly because of the fast growth rate of corneal cells. Acute effects of its absorption by the cornea are essentially pain and inflammation. It may feel as if sand is in your eyes. These symptoms are usually felt sometime between 30 minutes to 24 hours after the exposure, depending on its severity. More severe exposures are felt sooner. The worst of the symptoms are usually over after about 6 to 24 hours and rarely last longer than 48 hours, again because of the rapid cell growth rate. Very rarely does permanent damage result but also, unlike skin, the cornea does not develop increased tolerance with repeated exposure. The one type of permanent damage that can result to the cornea is through the growth of a pterygium over the surface of the eye with its associated asphericity.

UV-B is partially absorbed by the cornea (about 40%) and partially by the aqueous humor (about 10%.) It has similar effects on the cornea as UV-C. The rest of the photons (about 50%) are absorbed by the lens. The primary effect of repeated exposures of the lens is decreased transmission or increased scattering of visible light. The basic mechanisms or fundamental changes that result in this phenomenon, called cataract formation, are poorly understood. They are thought to be due to changes in cell structure, fluid imbalances, mineral buildups or protein aggregation. The damage is cumulative and permanent. Lens replacement is required to restore unobscured vision.

UV-A has less and less absorption by the cornea and aqueous humor as the wavelength increases toward visible light. So more and more is absorbed by the lens until the lens too starts to become transparent as the wavelength approaches visible light. As this occurs, the UV-A is passed onto the retina. This exposure can lead to accelerated aging of the retina, but far and away the primary damage again done by UV-A occurs in the lens as with UV-B.

Visible light can also cause damage to the eye and this will primarily occur in the retina. Because of the eye's sensitivity to visible light, protective reactions such as iris constriction, squinting or closing of the eye will generally protect it. If you persist in viewing bright visible light, you may suffer photochemical retinal injuries that can reduce the eyes ability to detect light.

Near infrared (700 to 1400 nm) is more dangerous than UV because a good portion of it penetrates through the various ocular media and is actually focused onto the retina. This results in heat buildup that can not be felt because there are no pain sensors there. The buildup of heat leads to denaturation of the biomolecules in the retina. In addition absorption by the other features such as the lens and the iris can also takes place. Here the lens is primarily damaged by tissue peeling off of the posterior surface of the lens. Damage to the iris is usually in the form of hemorrhages and inflammation.

Far infrared (1400 nm to 100 µm) is again primarily absorbed in the cornea. Delivered at low power it can again lead to itchy sore eyes. At moderate power, opacification caused by protein coagulation can result but pain felt on the skin usually causes you to close your eyes.

You have to understand that some eye damage is actually done by ambient levels of UV, visible and IR light all the time, but that this level is low enough that the natural on-going cellular renewal process is able to keep up with it. Something that upsets this balance, by decreasing the body's recuperative powers, can only make the situation worse. Chemicals or medicines that make the eye tissue more photosensitive can do this. Tetracyline, a common antibiotic is an example of a common drug that is a photosensitizer. You may also be interested to know though that individual or racial features do not seem to play as important role in eye sensitivity to light exposure as they do with the skin sensitivity. People with lens implants are also more sensitive to damage from light exposures.

There are a number of factors that dictate the severity of an optical exposure. They include: the temperature of the object, its size, its distance from your eyes and the length of the exposure.

The hotter an object is the more light that it emits and the wider the spectrum of the emitted radiation. Both of these are a result of the electrons getting raised to higher energy levels. This allows them to either have more small energy transitions as they fall to their ground state or have larger ones. More transitions mean more photons although of longer wavelength. Larger transitions means more dangerous light of shorter wavelength.

The size of the hot object defines the amount of material that is heated to high temperature. This translates directly to more electrons getting excited and thus more photons released. You get a large dose from peeking into a kiln because it is large.

The next factor that effects the exposure is the distance from it. The light from a kiln seems much brighter from 3 feet away than it does from across the room. This is because the density of photons decreases as you get further from the light source. Thus the farther away you are the fewer photons that hit your eye. The last factor, the length of exposure, can be minimized by keeping the amount of time in the kiln to a minimum.

Threshold limit value (TLV) standards have been set for UV, visible and IR light exposures by the American Conference of Governmental Industrial Hygienists (ACGIH). These standards are published as part of their annual booklet on TLV's for chemical materials in the workplace. These limits have been established to avoid injuries to the eye of the type listed above.

To determine whether the exposure you receive as calculated from the factors described above is bad or not you have to first evaluate the frequency distribution of that exposure. The radiation exposure is then weighted by the eye's sensitivity to that region and these products are summed over the three regions — UV, visible and IR. The weighted UV exposure should be less than 10 mW/cm^2 for exposures less than 1000 seconds.

The weighted visible light exposure should be below 100 kJ/m^{2}sr for exposure less than 10,000 seconds. Lastly, the time that the weighted IR exposure should be viewed is less than the square root of 10,000 divided by the product of the weighted exposure and the angle subtended on the eye. If these calculations are a little daunting don't feel alone.

It took Jim a long time to work this out, so if you don't understand it don't feel to bad. If at all possible, you should operate by the ALARA principle. Keep your optical radiation exposures As Low As Reasonably Achievable.

Eye protection

The basic type of protection from electromagnetic radiation for use by you in kiln casting of glass is absorbing filter glasses. One type of absorbing filter that you may already be aware of are welding filters. They are characterized by shade numbers where the higher the shade number the darker they are. They have the problem of making it hard to distinguish colors when wearing them; everything looks green, but this is not really be a problem in kiln work because all you are doing is topping off the glass anyway and everything is so hot they all look the same color. Another type of protection used by some kilnworkers is didymium glasses like those used by flameworkers. These are not really protective enough for kiln work.

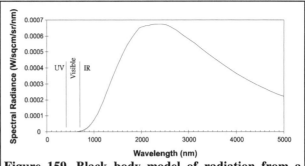

Figure 159. Black body model of radiation from a kiln.

So to investigate the effect of kiln peeking exposures Jim constructed a spreadsheet to calculate them. It first estimates the radiation spectra of the light coming out of the kiln by modeling it as black body radiation. From information in the scientific literature, this is a pretty good model except for the characteristic sodium flare. This distribution is illustrated in Figure 159. The graph shows the black body spectral distribution predicted for a kiln at 1700°F. From this figure you can see that almost all of the radiation given off from the kiln is in the IR portion of the spectrum.

Next this spectral distribution has to be weighted by the transmittance of whatever protective goggles are being used. The transmitted spectrum is then weighted by the eye sensitivity functions and compared to exposure standards.

For kiln work, we choose a test case of where a person is standing looking down into a partially open kiln at 1700°F. The kiln is 18 inches wide and open 10 inches high. The persons eyes are 30 inches from the open kiln. (Note that hotter kilns, larger openings or being closer to the eyes would all increase this exposure.)

Unprotected eyes would receive approximately a 315 mW/cm^2 exposure. This is more than 30 times the recommended exposure of 10 mW/cm^2. It could result in retina burns in as short an exposure as 24 seconds.

Rose didymium glass would cut this exposure down to about 76 mW/cm^2 but this could still result in retinal damage in as short as 26 seconds.

#4 or #5 green didymium shade would reduce this exposure down to about 2 mW/cm^2 and would allow viewing times up to about 7 minutes without damage. These shades are so dark though, that you will not be able to see very well what you are doing outside of the kiln. Therefore we recommend wearing a #4 shade as you look down into the kiln. AGW-99™ glasses by Aura Lens Products © would also be a good choice and they would still allow good color visibility.

Our recommendation is thus to wear glasses (not a face shield) that transmit only about 1% of the incident IR radiation. The advantage of this is that it allows your face to feel hot at about the same time as your eyes are starting to receive dangerous amounts of radiation. Any face and eye protection that you use should meet the standards of the American National Standards Institute's "Practice for Occupational and Educational Eye and Face Protection" (ANSI Z87.1).

Glossary

Alginate	a gelatin molding material made from algae that is useful for making short-term master molds.
Annealing point	temperature at which stress in glass is relieved in a few minutes (viscosity of 10^{13} poise)
Bas relief	a simple form of three dimensional sculpture which is like a plaque
Burnout	heating a mold up in the kiln to burn out any organic material
Calcine	process of heating up a substance to drive off water. Gypsum is calcined to make plaster.
Cire perdue	French term for lost wax casting
Coddle	a cylindrical mold frame constructed by wrapping a strip of flexible material like linoleum up
COE	coefficient of expansion of a glass
Colorant	material added to color glass - usually a metal oxide
Compatibility	ability to use two glasses in a casting without problems caused by their having differing coefficients of expansion
Compression	stress state where the atoms are closer together than they want to be
Consistency	amount of water used in preparing plaster—usually expressed in parts of water per 100 parts of plaster.
Covalent bonds	highly directional bonds where electrons are shared between two atoms
Density	property which is defined as the weight of an object divided by its volume
Devitrification	loss of vitreous (unorganized) structure of a glass
Dwell	section of a firing where the kiln temperature remains constant
Fining	process of allowing bubbles to rise to the surface of batched glass
Fins	casting defects similar to flashing on injection molded plastic parts that occur when the mold cracks.
Firing	a run in the kiln to process temperature and back
Flask	cylindrical metal mold frame used by jewelers
Flux	low melting-temperature enamel
Frit	crushed granules of glass
Fused silica	glass composed only of silica (silicon dioxide)
Fusing	process of joining pieces of glass together by softening them in a kiln
Gypsum	material (alabaster) from which plaster is made—chemically it is calcium sulphate ($CaSO_4 \cdot 2H_2O$)
Hard glass	glass that softens at a higher temperature
HEPA filter	high efficiency particulate filter
Insulation	a material with poor heat conduction
Investment	mixture of materials from which refractory molds are cast
Ionic bonds	non-directional bonds consisting of one negatively charged and one positively charged ion held together by electrical forces
IR	light with wavelengths just longer than visible light
Jiffy mixer	bladed attachment used with a drill to power mix investment
Kevlar	low expansion form of fiberglass used to make heat resistant gloves
Kiln	heated chamber, usually electrically, use to cast glass

© 2000 James Kervin and Dan Fenton

Lime	calcium carbonate ($CaCO_3$) which decomposes to calcium oxide (CaO) by release of carbon dioxide during batching
Master mold	a mold used to cast multiple wax replicas of your model
Mesh number	measure of the size of particles based on the size of particles that pass through a wire mesh. Meshes are numbered by the number of openings per linear inch
Mold	a spatial negative of your model in which a casting is made
Mold frame	container in which the refractory mold is invested
Natch	a locking or keying feature added to allow easy registry of multi-part molds
Opacity	inability of a glass to pass an image through it
Palette	base upon which model is constructed and mold frame is anchored to during investment
Pâte de Verre	literally translated from French as "paste of glass"
Peaking	point in hand mixing of investment where little islands or peaks form above the water surface.
Plastecine	oil-based clay with properties almost like a wax. (Will not mix well with waxes.)
Poise	a measure of viscosity
Ramp	section of the firing where the kiln is changing temperature by either heating or cooling - usually at a constant temperature
Refractory	a material resistant to high temperatures and attack by materials at high temperatures
Replica	a duplicate model cast in a master mold
Reservoir	cup-like addition to the top of a mold to hold extra glass needed to fill casting as glass frit consolidates
Screeding	scraping the top of an investment pour with a flat instrument to get a flat-bottomed mold
Separator	material used to coat a model or mold frame to make removal easier
Slaking	process of plaster combining physically with water. A few minute time period allowed after adding plaster to water but before mixing.
Soda	sodium carbonate (Na_2CO_3) which decomposes to sodium oxide (Na_2O) by release of carbon dioxide during batching
Soda-lime glass	glass whose composition is based on silica, soda and lime
Soft glass	glass that softens at a lower temperature
Softening point	temperature at which glass sags quickly under its own weight (viscosity of 4×10^7 poise)
Strain	differences in length between two marks caused by stress (expressed in micro inches/inch)
Strain point	temperature at which stress in glass is relieved in a few hours (viscosity of 3×10^{15} poise)
Stress	interior forces in the glass resulting from differences in strain or thermal gradients
Tap density	density achieved by gently tapping a container of frit on a table
Tension	stress state where the atoms are further apart than they want to be
Tetrahedron	four sided solid with triangular shaped sides
Thermal shock	fracture of the glass in a kiln by thermal gradient induced stresses
Translucency	measure of a the amount of light a casting transmits through it
UV	light with wavelengths just shorter than visible light
Viscosity	a material's resistance to flow
Working point	temperature at which glass can be manipulated and formed by hand (viscosity of 10^4 poise)

Suppliers and Manufacturer Sources

A. P. Green Industries Inc.
Green Boulevard
Mexico, MO 6525
Tel (573) 473-3626   Fax (573) 473-3495
Manufacturer of a complete line of refractory products including firebrick, ceramic fibers, insulation and castable refractory mixes.

Arrow Springs
4301 A Product Drive
Shingle Springs, CA 95682
Tel (530) 677-1400   Fax (530) 677-1600
www.arrowsprings.com
Distributor of Ransom and Randolph investments, wax working tools and Bullseye frit. Manufactures kiln and tools.

A.R.T.C.O.
348 N. 15th St.
San Jose, CA 95112
Tel (408) 228-7978
Manufacturer of glassblowing supplies. Carry an assortment of high temperature gloves; Kevlar, Kevlar/PBI blend and pure PBI.

Carborandum, Fiberfrax Insulation
Standard Oil, Engineered Materials Co. Fibers Division
P. O. Box 808
Niagara Falls, NY 14302
Tel (716) 278-2183
Manufacturer of alumina-silicate based fibrous ceramic insulation materials such as bulk fiber, fiber blanket, papers, textiles, coatings and boards.

Bullseye Glass Co.
3722 SE 21st Ave
Portland, OR 97202
Tel (503) 232-8887   Fax (503) 238-9963
Manufacturer of compatible colored sheet glass and frit (CTE 90) in opals, transparents and some dichroic. Will not sell direct, but they will advise you as to the best source of supply for your area. Publish a newsletter on their products and program.

Ceramic Craft and Supply Co.
490 5th St.
San Francisco, CA 94103
(415) 982-9321
They had Pemco PB83 at the time of this writing.

Columbus Dental
1000 Chouteau Ave.
(P.O. Box 620)
St. Louis, MO 63188
(314) 241-2988
Distributor of dental investment plasters.

C & R Loo
1085 Essex
Richmond, CA 94804
Tel (800) 227-1780   Fax (510) 232-7810
www.crloo.com
Distributor of colored frit from Germany. Also Bullseye frit and Lenox crystal. Proof of being a wholesale business is required.

Crystalite Corporation
8400 Green Meadows Dr.
Westerville, OH 43081
Tel (614)548-4855   (800)777-2894
FAX (614)548-5673   www.crystalite.com
Manufacturer of diamond tooling for lapidary and glass market. Large wheels to small bits.

Denver Glass Machinery
2800 South Shoshone St.
Englewood, CO 80110
Tel (303) 781-0980   Fax (303) 781-9067
www.denverglass.com
Manufacturer of professional coldworking equipment including, disk grinders, spindle grinders, bevelers, etc. Distributor of 3M and Roloc sanding disks.

Digitry Co. Inc.
188 State St. Suite 21
Portland, ME 04101
Tel (207) 774-0300   Fax (207) 774-3291
Manufacturer of multi-ramp programmable controller for kilns.

Douglas and Sturgess
730 Bryant St.
San Francisco, CA 94107
(415)896-6383   (888)278-7883
FAX (415)896-6379
Distributor model and mold making supplies such as waxes, latex, plaster, moulage, etc..

Dremel
P. O. Box 1468 Dept 57
Racine, WI 53406-1268
www.dremel.com
Manufacturer of compact power tools used for grinding available at hardware and hobby stores.

Ed Hoy's
27625 Diehl Rd.
Warrenville, IL 60555-3838
Tel (800) 323-5668   (630)836-1353
Fax (708) 416-0448   www.edhoy.com
Distributor of complete hot and cold glassworking equipment and supplies. They have crucible, frits, glasses, investment mixes, Zetex gloves, etc. Strictly wholesale.

Edmund Scientific
101 E. Glouecester Pike
Barrington, NJ 08007-1380
(607) 573-6250   www.edsci.com
Distributor of polarizing filters for testing compatibility as well as a multitude of scientific curiosities.

Fenton Glass Studio
851 81st Ave
Oakland, CA 94621
(510) 638-1313   Fax (510) 638-1314
Classes in many glassworking topics, which include: pâte de verre, fusing, enamels, sandblasting, stained glass, beadmaking, etc.

Frei & Borel
P. O. Box 796
Oakland, CA 94604
Tel (800) 772-3452   Fax (800) 900-3734
http://paleoart.com/frei/index.htm
Distributor of jewelry making supplies and including waxes and wax working equipment

Fritworks
P.O. Box 158
Zieglerville, PA 19492-0158
(888) FRTWRKS (610) 287-0221
Fritwrks@gateway.net
Distributor of Bullseye frit, molds and other supplies For kiln casting and pâte de verre work.

Georgia Pacific
Gypsum Division
900 South West 5th Avenue
Portland, OR 97202
(800) 225-6119   www.gp.com
Manufacturer of many kinds of gypsum plasters and cements.

Glass Fusion Studio
1618 S. E. Ogden St
Portland, OR 97202
(503) 235-2284
Distributor of hot glassworking equipment and supplies.

His Glassworks Inc.
91 Webb Cove Rd.
Asheville, NC 28804
(800)91-GRIND   (800) 914-7463   (828) 254-2559
Fax (800)254-2581   www.hisglassworks.com
Distributor to glass artists of grinding equipment and Crystalite diamond tools.

Jiffy Mixer Co. Inc.
4120 Tigris Way
Riverside, CA 92503-4843
(800)5602903 (909)272-0838 FAX (800)666-4120
Manufacturer of efficient mixing heads that work well for power mixing of investment it minimizes splashing and air entrapment.

Lapcraft Company
195 West Olentangy St.
Powell, OH 43065
(614) 764-8993
Manufacturer of diamond grinding bits available at many lapidary shops.

Lenox Crystal
Mt. Pleasant, PA 15666
(412)547-4541
Manufacturer of clear lead crystal cullet. Will deal direct.

OMEGA Engineering, Inc.
P.O. Box 4047
Stamford, CT 06907-0047
(800) 848-4286 (203) 359-1660 Fax (800) 622-2378
www.omega.com
Manufacturer of temperature measurement and control systems. Call number to order catalogs. Fax number for customer service.

Supplier and Manufacturer Sources

The Edward Orton Jr. Ceramics Foundation
6991 Old 3C Highway
Westerville, OH 43081
(800) 999-5442   FAX (614) 895-5610
Manufacturer of pyrometric cones, kilns and kiln venting systems

Pacific Glass
125 W. 157th St.
Gardina, CA 90248
(800) 421-5170 (310) 576-7823 FAX (310) 516-0335
California (800)354-5277   www.pacificglass.com
Distributor of Bullseye frit and fusing supplies including crucibles. Wholesale only so you need to qualify.

Polytek Development Corp.
55 Hilton St.
Easton, PA 18042
(610) 559-8620   Fax (610) 559-8626
www.polytek.com
Manufacturer of all kinds of flexible molding materials including polyurethane, silicones, latex's, & alginates.

Premier Wax Co.
3325 Hidden Valley Dr.
Little Rock, AR 72212
(501) 225-2925
Manufacturer of a variety of different waxes

Ransom and Randolph Div.
120 W. Wayne St.
Maumee, OH 43537
(419) 893-9497   Fax (419) 893-4988
www.ransom-randolph.com
Manufacturer of a number of different castable refractory materials. Will sell direct and ship from nearest warehouse.

RIBTEC
Ribbon Technology Corporation
6270 Bowen Rd.
Canal Winchester, OH 43110
(800) 848-0477
Manufacturer of stainless steel fiber reinforcement.

Rio Grande Jewelry Supply
6901 Washington N.E.
Albuquerque, NM 87109
(800) 545-6566   Fax (800) 965-2329
Distributor of jewelry supplies including waxes and wax working equipment

Standard Ceramic Supply
P. O. Box 4435
Pittsburgh, PA 15205-0435
(412)276-6333   Fax (412)276-7124
www.standardceramic.com/clay/
Distributor of ceramic supplies including lead based frits. We recommend # 28 which is 65% PbO, 1% Al_2O_3 and 34% SiO_2.

Steinert
1507 Franklin Ave.
Kent, OH 44240
(800) 727-7473 (330) 678-0028 Fax (330) 678-8238
Manufacturer of the glass crusher.

Thompson Enamels
P. O. Box 310
Newport, KY 41072
(606) 291-3800
Manufacturer of enamels and fine frits as well as Kylnfire binder. They will sell direct, although they may have distributors closer.

Trinity Ceramic Company
9016 Diplomacy Row
Dallas, TX 75247
(214) 631-0540
Distributor of ceramic supplies including Pemco PB 83 lead glass frit.

Truebite, Inc.
2590 Glenwood Rd.
Vestal, NY 13850-2936
(800)-676-8907   Fax (607) 785-2405
Distributor of small tools to the crafts industry. Carry diamond and silicon carbide bits, sanding disks, and cut-off wheels for moto-tools. Also carry small hand tools like diamond pads, files and points.

United States Gypsum Company
Industrial Gypsum
125 South Franklin
Chicago, IL 60606-4678
(800) 874-4968   (312)606-400
www.usg.com
Manufacturer of gypsum plasters and cements.

Uroboros Glass Studios
2139 N. Kerby
Portland, OR 97227
(503) 284-4900   www.uroboros.com
Manufacturer of expansion 90 & 96 colored sheet glass mostly in dichroics. Will not sell direct but will advise. Publish a newsletter.

© 2000 James Kervin and Dan Fenton

Wale Apparatus
400 Front St.
Hellertown, PA 18055
(800) 334-9253   FAX (610)838-7440
Distributor of protective gloves and glasses, and grinding equipment.

Wisconsin Aluminum Foundry Co. Inc.
836 S. 16th St., P.O. Box 246
Manitowoc, WI 54220
Tel (414)682-8286 Fax (414)682-7285
Manufacturer of industrial steam systems as well as heavy cast aluminum pressure cookers/canners that can be modified for steam systems. The 21 ½ quart size is about $125 and the 15 ½ one is about $110.

Zircar Products Inc.
P.O. Box 287
87 Meadow Rd.
Florida, NY 10921
Tel (914) 651-3040   Fax (914) 651-0074
www.zircarzirconia.com
Manufacturer of fibrous ceramic insulation materials such as alumina-silicate and zirconia fibers.

References

Algeren, Ray. **Fusing with Spectrum Glass**, Spectrum Glass Company, 1986 *This short paper gives the only instruction on how to perform the bar test for compatibility. It is out of print, but if you call the company they might send you a Xerox.*

Ammen, C. W. **The Metalcaster's Bible**, Tab Books Inc., 1980. ISBN 0-8306-1173-8 *A basic book on casting of metal. Discusses mold making, core molds and mold making tools. It also has an extensive dictionary of casting terminology.*

Barrie, Bruner Felton. **Mold Making, Casting & Patina for the student sculptor**, Adams, Barrie, Felton and Scott Publishing, 1992. ISBN 0-9631867-0-1 *A nice short reference book on plaster, silicone rubber and moulage mold making for plaster castings.*

Bloch-Dermant, Janine. **G. Argy-Rousseau Glassware as Art**, Thames and Hudson, 1991. ISBN 0-500-23626-7 *A beautiful book with lots of delightful pictures of Pâte de Verre by G. Argy-Rousseau with some description of his techniques.*

Bovin, Murray. **Centrifugal or Lost Wax Casting Jewelry Casting for Schools Tradesmen Craftsmen**, Bovin Publishing, 1977. ISBN 0-910280-05-3 *Basic text on metal casting. Good set of conversion tables in the back of book.*

Bray, Charles. **Dictionary of Glass, Materials and Techniques**, University of Pennsylvania Press, 1995. ISBN 0-7136-4008-1 *A good general purpose encyclopedia-like reference that everyone working in glass should have.*

Chaney, Charles and Skee, Stanley. **Plaster Mold and Model Making**, Van Nostrand Reinhold Co., 1973. ISBN 0-671-60896-7 *This one is in print as a paperback and is available at most ceramic suppliers. The chapter on mixing plaster alone makes it worth it. It goes into two-part molds and then on to multi-part molds.*

Clark, Nancy; Cutter, Thomas and McGrane, Jean-Anne. **Ventilation: A Practical Guide for Artists, Craftspeople, and Others in the Arts**, Lyons & Burford, 1984. ISBN 0-941130-44-4 *Gives a good basic description of ventilation systems with examples. It also provides much of the information needed to size your own system.*

Clayton, Pierce. **The Clay Lover's Guide to Making Molds, Lark Books**, 1998. ISBN 1-57990-022-4 *A good text on the making of plaster molds for slip casting and press molding of clay.*

Coburn, Andrew, Dudley, Eric and Spence, Robin. **Gypsum Plaster Its manufacture and use**, Intermediate Technology Publications, 1989. ISBN 1-85339-038-0 *A short book on gypsum plaster—its chemistry, origin and manufacture.*

Cummings, K. **The Technique of Glass Forming**, B. T. Batsford (London), 1980. ISBN 0-7134-1612-2 *Printed by the Anchor Press Ltd., 4 Fitzhardinge St, London, W1H0AH. Encouragement to experiment with kiln working. It was out of print almost before it was published. A must for the book collector but keep in mind that much of the material is theoretical.*

Cummings, K. **Techniques of Kiln-formed Glass**, A & C Black Limited (London), 1997. ISBN 0-8122-3402-2 *Just as good as the previous book with much of the same information in case you missed the other. Get it.*

Denol, Edutions, **La Pâte De Verre**, 1984, ISBN: 2.207.1006.7. *You might have to go to France to get this one. It is written in French. Good photos of late 1890's European work. The best of pâte de verre, but we can't read it.*

Eliscu, Frank. **Sculpture Techniques in Clay, Wax and Slate**, Chilton Company, 1959. *Out of print. Gives some basic information on sculpting with these materials.*

Fielder, Larry. "Mold Casting of Glass" **Hot Glass Information Exchange 1979**, John Bingham ed., 1979. *This is an oldie and only 1,000 copies were printed but reprints are available from Whitehose Books.com. These are some of Larry Fielder's first writings on pâte de verre. Collectible if you're a glassomaniac.*

Fielder, Larry. "Pate de Verre, adapted from a Larry Fielder workshop." **Glass Art**, Vol. 6 No. 6 (Sept/Oct 1991). *These are some of Larry's later writings on the subject of pâte de verre. You should still be able to get copies of it.*

Frith, Donald E. **Mold Making for Ceramics**, Chilton Book Company, 1985. ISBN 0-8019-7359-7 *A bit pricy but a good reference text on slip casting mold manufacture using plaster.*

Fufimori Inc, **Pâte de Verre Kimiake Higuchi and Shinichi Higuchi Selected Works**, Published by Ryutaro Adachi 1997. ISBN 4-7630-9732-6 *A beautiful book of the some of the works of Kimiake and Shinichi Higuchi. Inspirational for all*

Gardner, Paul V. **Frederick Carder: Portrait of a Glass maker**, The Corning Museum of Glass, 1985. ISBN 0-87290-111-4 *A good book showing a cross section of the life's work of Frederick Carder including some of his kiln cast work. Limited discussion of techniques.*

Halem, Henry. **Glass Notes — A Reference for the Glass Artist**, Third Edition 1996, ISBN 1-885663-02-1 Halem Studios, Inc., 429 Carthage Ave., Kent, OH 44240-2303 Tel (216) 673-8632, *Contains a wealth of information about glass, casting, equipment construction, etc.*

Hollister, Paul. "Pâte de Verre: The French Connection." **American Craft**, Vol. 48 No. 4 (Aug/Sept 1988). *A description of the rediscovery of Pâte de Verre and work of around the turn of the century.*

Kallenberg, Lawrence. **Modeling in Wax for Jewelry and Sculpture**, Chilton Book Co. 1981. ISBN 0-8019-6896-8 *Excellent book on the modeling techniques used with wax by jewelers. Good discussions on the making of multiples using RTV.*

Kenny, John B. **The Complete Book of Pottery Making**, Chilton Book Company. First printing in 1949. *May or may not still be in print. A really good book on working with plaster and making molds of all configurations.*

Kervin, James E. **More Than You Ever Wanted To Know About Glass Beadmaking 4th ed.**, GlassWear Studios, 2000. *In print and available. Has a chapter on how the Pâte de Verre process can be used to make glass beads. It's chapter on fused glass beads is also applicable.*

Kohler, Lucartha. **Glass, An Artist's Medium**, Krause Publications 1998. ISBN 0-87431-604-x. *A gook book covering a wide variety of glass related topics including kiln casting.*

Lillie, H. R., "Basic Principles of Glass Annealing" **The Glass Industry** Vol. 31 No.7 (July 1950) *Formed the basis for the classical understanding of glass annealing. Not for the weak of math.*

Supplier and Manufacturer Sources

Lundstrom, Boyce and Schwoerer, D. **Glass Fusing - Book One**, Vitreous Publications, 1983. ISBN 0-9612282-0-2 *It is still in print and is probably the best book available on basic fusing. An excellent chapter on testing for compatibility. Also a good basic overview of kiln firing procedures and glass annealing.*

Lundstrom, Boyce. **Advanced Glass Fusing - Book Two**, Vitreous Publications, 1989. ISBN 0-961228-1-0 *Still in print. Good chapter on bas relief glass casting.*

Lundstrum, Boyce. **Glass Casting and Moldmaking - Glass Fusing Book Three**, Vitreous Publications, 1989. ISBN 0-9612282-2-9 *In print and available. It is worth it. Discussions on pâte de verre, frit casting and mold making.*

McCann, Michael. **Health Hazards Manual for Artists**, Nick Lyons Books, 1985. ISBN 0-8230-0295-0 *Basic discussion of many of the hazards associated with different art forms and how these materials can affect your body.*

McCreight, Tim. **Practical Casting A Studio Refereence Revised Edition**, Brynmorgen Press, 1994. ISBN 0-9615984-5-x *Good practical book on metal casting. Jam packed with information on lost wax casting.*

McLellan, George W and Shand, E. B. **Glass Engineering Handbook 3rd ed.**, McGraw-Hill Book Company, 1984. ISBN 0-07-044823-X *Lots of technical information on glass properties and processes. In print. Not for math weenies or faint of chart.*

Midgley, Barry. **The Complete Guide to Sculpture, Modeling and Ceramics Techniques and Materials**, Chartwell Books, Inc., 1986. ISBN 0-89009-471-3 *Nice book on working with these materials. Lots of good pictures.*

Miller, Richard McDermott. **Figure Sculpture in Wax and Plaster**, Watson-Guptill Publications, 1971. ISBN 0-8230-1720-6 *In print in a paperback edition by Dover last we looked. This is one of the better books on describing how to go about doing sculptural work with wax. It also discusses preparing for lost wax casting. Its description of plaster work also gives one a feeling for that material and how molds can be built up around large pieces.*

Morman, Shar. **Warm Glass**, CKE Publications, 1989. *In print and available. Very comprehensive in all aspects of kilnwork. She also addresses the mold and plug method of casting vessels. There were some technical glitches in the first edition so make sure that you get the addendum when you buy the book.*

Narayanaswamy, O. S. "Annealing of Glass" in **Glass Science and Technology** Edited by D. R. Uhlmann and N. J. Kreidl **Volume 3 Viscosity and Relaxation**, Academic Press, Inc. 1985. ISBN 0-12-706703-5 *A good summary of the modern theory of annealing. This book puts even Jim's grasp of mathematics and thermodynamics to the test. We suggest a Bachelor's in Engineering before you even attempt reading this one.*

Reynolds, Gil. **The Fused Glass Handbook**, Hidden Valley Books, 1987. ISBN 0-915807-02-5 *In print. Good introduction to testing for compatibility. He has written more detailed articles since the book.*

Rhodes, Daniel. **Clay and Glazes for the Potter**, Chilton Book Company, 1971 ISBN 0-8019-0165-0 *The definitive work on clays and glazes. Good information on colors and composition of glazes which are really glasses.*

Rosenblatt, Sidney. "The 'Lost' Art of Pâte de Verre" **Hobbies**, Vol. 73, No. 8 (Oct 1968). *A short general introduction to the pâte de verre process.*

Rossol, Monona. **The Artist's Complete Health and Safety Guide**, Allworth Press, 1990. ISBN 0-927629-10-0 *Another basic discussion of many of the hazards with different art forms and how these materials can affect your body. A little information on the current practices and regulations.*

Scholes, Samuel R. **Modern Glass Practice**, c 1935 ISBN 1-878907-07-7 7th revised edition, CBI Publishing Company. *This is a highly technical engineering manual. Much of this book is difficult to understand. The first chapter will give you an understanding of glass as well as its material properties. There is also an excellent chapter on the classical theory of annealing.*

Schuler, Fredrick and Lilli, **Glassforming**, Chilton Book Company, 1970. ISBN 0-8019-5558-0 *Out of print. Has an interesting chapter on annealing. It also has a small section on mold and plug casting for forming bowl-shaped vessels.*

Tokyo Glass Art Institute, **The Art and Technique of Pâte de Verre**, 1998. ISBN 4-9980613-2-1 *The book is an excellent text on the process of pâte de verre. It has great pictures on the process and brief right-to-the-point commentary. A must for every kiln casting artist's shelf.*

Weinberg, Steven. "Glass Casting Techniques" **Hot Glass Information Exchange 1979**, John Bingham ed., 1979. *Out of print but reprints are available from Whitehose Books.com. Discussion of the basic casting process.*

Weyl, Woldemar A. **Coloured Glasses**, Society of Glass Technology, Sheffield, England 1951. *In its 5th printing, it is still one of best books around on the coloring of glass. It is fairly technical though.*

Index

A.F.S. number, 79
abrasive cleaners, 151
acetate sheet
   use in making coddles, 86
acrylic
   use in making multi-part freehand molds, 99
   use in preparing clay models, 56
advance casting techniques, 109
air drying
   reason for doing to molds, 19
alginate
   definition of, 169
   description of, 64
   using to make master molds, 64
alumel, 134
alumina hydrate
   as a refractory, 78
alumina-silicate fibers
   as a property modifier, 79
annealing
   allowable residual stress levels, 38
   annealing region, 37
   calculating of allowable cooling rates, 37
   definition of, 36
   effect of casting shapes on cooling rates, 130
   factors in cooling, 37
   studio determinion of properties, 40
   theory of, 36
annealing point
   definition of, 169
   determination of, 37
   explanation of, 32
   studio determination of, 40
asbestos
   use in high-temperature gloves, 114
backhand flick
   use in making freehand molds, 99
band saws, 138
   use of, 138
bas relief molds
   definition of, 169
   using fiber blanket to more uniformly cool, 131
belt sanders, 147
   use of, 147
bench peg, 49
binders for investments
   cements, 77
   chemistry of, 74
   colloidal solutions, 77
   description of, 73
   gypsum plasters and cements, 74
   list of, 74
   mechanisms of, 74
binding agents
   for glass pastes, 116
bits
   for hand-held grinders, 145
Blazer Piezo, 52
blenders
   use in frit manufacture, 113
Bon Ami, 151
bonds
   covalent, 26
   ionic, 26
bridging oxygen atoms, 26
brushing, 150, 151
   wheels for, 151
bubbles
   effect on opacity, 11
   minimizing, 110
   removal from investments, 17
   result of, 109
buddy system
   use in crucible handling, 114
buffing, 149
   mounting of wheels, 151
   preparation of wheels, 150
   use of wheels, 150
   wheels for, 150
Bullseye Glass, 28
   allowable cooling rates for, 38
   available frit sizes of, 116
   COE of, 33
   leachability of, 30
burnout, 121
   definition of, 169

fumes from, 20
temperatures for, 122
burns
   degrees of, 157
   treatment of, 158
   treatment of chemical, 159
   treatment of chemical ones, 159
   treatment of thermal one, 159
buttresses
   using to reinforce mold frames, 14
calcine
   definition of, 169
canners
   using to steam wax out of molds, 106
casting waxes
   injection of, 70
   use in replica casting, 70
castings
   hand cleaning of, 22
   natural finish of, 23
   removal from mold, 22
cements
   calcium alumina cement, 77
   Fondu, 77
chemically bound water, 121
chicken wire
   using for internal reinforcement of molds, 88
chipping
   during grinding, 146
   when sawing, 141
chisels
   use in carving plaster, 57
chronic exposures, 162
chrysolite, 79
cire perdue, 43
   definition of, 169
clay
   coiling of, 54
   composition of, 53
   desirable properties of, 53
   handworking of, 55
   kind used for modeling, 54
   modeling with, 52
   origin of, 52
   pinching of, 54
   plastecine, 46
   preparing model for investment, 56
   press modeling with, 55
   primary, 53
   removal from a mold, 19
   secondary, 53
   slabbing of, 55
   use in making multi-part molds, 90
   using to cast wax shapes, 48
   using to make master molds, 64
   working with, 54
CMC
   use as a binding agent, 116
coddles
   construction of, 86
   definition of, 169

COE
   definition of, 169
   explanation of, 31
   measurement of, 31
   range of, 33
   table of, 33
   using to calculate size changes, 33
   variability of, 31
coiling
   in clay modeling, 54
coldworking, 137
colloidal alumina, 77
colloidal silica, 77
color
   blending of during casting, 110
   of glass during firing, 21
colorants, 40
   definition of, 169
   factors affecting, 41
   forms of, 40
   interaction of, 110
   recipies of, 41
   safety practices in handling of, 41
   use in making colored frit, 113
   using to color glass, 41
compatibility
   COE requirements for, 33
   definition of, 169
   explanation of, 33
   factory testing of, 34
   high temperature requirements for, 33
   sudio testing of, 34
   Transitive Property of Glass Fusing, 34
   working around incompatibility, 35
compression
   definition of, 169
cones
   using, 135
   using to calibrate thermocouples, 135
consistnecy
   definition of, 169
constantin, 134
cooling phase, 122
cooling rates
   calculating of allowable cooling rates, 37
   effect of casting shape on, 38
   effect of mold thickness on, 39
   factors in allowable, 37
cope
   part of hollow vessel molds, 96
covalent bonds
   definition of, 169
cracks in molds
   repairing, 103
cradle molds
   using for external reinforcement of molds, 89
   using to secure multi-part molds, 91
crashing
   procedures for, 129
crucibles
   handling of, 113

safe handling of, 114
use in drip casting, 110
use in frit production, 112
crystobalite
    as a refractory, 78
curing
    considerations in, 121
    of molds, 19, 121
    of molds, 18
    temperatures for, 121
cutoff saws, 139
cuts
    treatment of, 155
cystalline
    structure of, 26
deadman's switch, 114
deaerating
    of investments, 86
density
    definition of, 169
density ratios
    table of ratios, 111
    using to determine amount of frit required, 111
dental tools
    using on waxes, 49
devitrification
    definition of, 169
    location of, 27
    temperature range of, 21
dextrine
    use as a binding agent, 116
diamond embedded tools, 137
    diamond bonding, 138
    files, 149
    grit definition, 137
    grit size use, 138
    pads, 149
diatomite
    as a refractory, 78
didymium glasses, 167
disk grinders, 142
    construction of, 142
    safe use of, 143
    use of, 143
disk sanders, 147
draft
    definition of, 18
drag
    part of hollow vessel molds, 96
drip casting, 110
drum sanders, 148
    types of, 148
drums
    using to steam wax out of molds, 106
drying
    increasing speed of, 107
    of molds, 107
dwell
    definition of, 169
electromagnetic radiation, 164
    exposure factors, 166

eye damage from, 165
IR, 165
protection from, 167
protective glasses for, 167
sensitation to, 166
spectrum of, 164
TLV for, 167
UV, 165
visible light, 165
Elmers glue
    use as a binding agent, 116
enamels
    use in pastes, 117
end-point control
    in programmable controllers, 133
etching compounds, 152
extruder
    for wax wires, 48
Ferro 3419-2, 29
fiberglass fibers
    as a property modifier, 79
files
    cleaning of, 50
    types for waxes, 50
filing
    of waxes, 50
filling a mold at temperature, 124
fining, 11
    definition of, 169
finishing techniques, 137
    buffing and brushing, 149
    cleaning and etching, 151
    coatings, 152
    cutting, 138
    for replica castings, 72
    general steps, 137
    grinding, 141
    sandblasting, 152
    sanding, 147
    sawing, 138
fins
    definition of, 169
fire extinguishers
    operation of, 157
firebricks
    using for external reinforcement of molds, 89
fires
    types of, 157
    what to do in case of, 157
firing, 122
    considerations in, 122
    definition of, 169
    diagnosing too high a temperature of, 21
    phases of, 122
    visual checking progress of, 21
flame polishing of waxes, 52
flammable liquids
    safe storage of, 156
flasks
    definition of, 13, 169
    using for external reinforcement of molds, 88

flow
    using to tell firing progress, 21
flowerpots
    use in drip casting, 110
flux
    definition of, 169
foils
    use as inclusions, 120
Fondu, 77
frit
    availability in Bullseye glass, 116
    availability in Uroborus glass, 116
    coloring of, 113
    commercial sizes of, 116
    definition of, 169
    determing amount required, 111
    hand crushing, 12
    manufacture of, 111
    packing of, 30
    pour density of, 31
    preparation of, 11
    reasons to mix types of, 28
    shrinkage in casting, 12
    size measurement of, 30
    sizing of, 12
    tap density of, 31
    typical mesh sizes used, 30
frit production
    mecanical methods of, 112
    thermal methods of, 112
fused silica
    definition of, 25, 169
    melting temperature of, 26
    properties of, 27
fusing
    definition of, 169
fusing test strip
    edge frit version, 35
    edge strip configuration, 35
    marking of, 34
    reading of, 34
    running of, 34
    using interference filters with, 34
garbage disposals
    use in frit manufacture, 113
gating
    in press molds, 94
    of models, 61
gelatin
    use as a binding agent, 117
Gelflex
    description of, 65
    types of, 65
    using to make master molds, 65
GFI, 86
glass crusher
    commercial, 112
glass pastes. See pastes
glasses
    annealing of, 36
    annealing point of, 32
    chemistry of, 25
    choosing a pallete of, 25
    coloring of, 113
    compatibility of, 33
    effect of iron on color, 29
    hard, 26
    lowering melting temperature of, 26
    physical forms for casting, 25
    soft, 26
    softening point of, 32
    strain point of, 31
    structure of, 25
    table of types, 27
    temperature regimes of, 31
    types of, 27
    viscosity changes of, 27
    working point of, 32
glazes
    production of, 115
glossing off
    of investments, 18
glovebox, 160
gloves
    chemical resistance of, 158
    types of high-temperature ones, 114
glue guns
    safety with, 14
    sticks for, 14
    use in construction of mold frames, 14
GNA, 28
    annealing and strain point of, 40
grinding
    grit order, 141
grinding equipment, 142
    disk grinders, 142
    hand-held grinders, 145
    spindle grinders, 143
    use of, 146
grinding wheels
    care of, 144
    types of, 144
    use of, 144
grits
    sizes of, 137
    sizes order of use, 138
grog
    as a property modifier, 79
ground fault interrupt, 86
gum arabic
    use as a binding agent, 116
gum tragacanth
    use as a binding aget, 116
gypsum
    definition of, 169
gypsum cements
    Ultracal, 77
    USG Hydrocal, 76
    USG Hydroperm, 76
    USG Hydrostone, 77
gypsum plasters
    description of, 74

Index

determining usability of, 76
list of properties, 75
modifying performance of, 75
normal consistency of, 75
storage of, 74
strength of, 75
USG #1 casting plaster, 76
USG #1 molding plaster, 76
USG #1 pottery plaster, 76
hairdrier
   using to dry molds, 107
hammer
   use in frit manufacture, 112
hand sanding, 149
hand-held grinders, 145
   accessories for, 145
   bits for, 145
   use of, 145
handsaws, 140
handworking
   of clay, 55
   of waxes, 51
hard glass
   definition of, 169
heating phase, 122, 123
   effect of mold insulation on, 123
HEPA filter
   definition of, 169
hollow vessel molds
   construction of, 97
   cope part of, 96
   description of, 96
   drag part of, 96
   sectional technique, 97
   using packing instead of a drag, 98
   vents for, 97
Hydrocal cement, 76
Hydroperm cement, 76
Hydrostone cement, 77
inclusions
   glass, 119
   investment, 120
   lampworked, 119
   metallic, 120
   negative space, 120
   types of, 119
infinite range switches, 131
infrared radiation, 165
insulation
   definition of, 169
interference filters, 34
   appearance of stress levels, 35
   using to measure stress, 34
investment inclusions
   manufacture of, 120
   removal of, 120
   to create negative space images, 120
investments
   acceleration of
      effect of dirty tools, 16
      hot water, 16
      stirring, 17
   adding more of, 17
   applying first layer of, 17
   binder component, 73
   choosing between premades and self-made, 73
   commercially available mixes, 83
   component materials, 73
   deaerating, 86
   definition of, 169
   disposal of, 17
   effect of mix ratios on, 81
   formulation of, 80
   glossing of, 18
   hazards of, 80
   how much to mix, 17
   maximum solubility temperature of, 16
   mix # 1, 81
   mix # 2, 82
   mix # 3, 82
   mix # 4, 82
   mix # 5, 82
   mix # 6, 82
   mix # 7, 83
   mix # 8, 83
   mixing large batches of, 85
   mixing of, 15
   origin of term, 73
   pouring of, 17
   property modifiers, 79
   refractory component, 77
   removing bubbles from, 17
   safety with, 15
   screeding of, 18
   setting of, 18
   slaking of, 16
   stirring of, 16
   using to secure multi-part molds, 92
ionic bonds
   definition of, 169
IR
   definition of, 169
Jiffy mixer, 85
   definition of, 169
jigs
   use in sawing, 141
kaolin clay
   as a property modifier, 80
   as a refractory, 78
   origin of, 53
Kevlar
   definition of, 169
   use in high-temperature gloves, 114
kiln controllers, 131
   kiln sitters, 132
   programmable controllers, 133
   set point controllers, 132
   types of, 131
kiln operations
   crashing procedures for, 129
   devitrification during, 128
   effect of mold wetness on, 125

effect of wax steam out on, 125
heating and cooling rate considerations, 123
kiln controllers for, 131
proper dress for, 125
safety in, 126
segment 10—dwell to allow equilibration after crash cooling, 129
segment 11—controlled cooling through annealing zone, 129
segment 12—dwell after cooling through annealing zone, 130
segment 13—final cooling segment, 130
segment 1—dwell to remove chemical water, 124
segment 1—initial heating segment, 123
segment 2—dwell to remove physical water, 123
segment 3—second heating segment, 124
segment 5—third heating segment, 124
segment 6—dwell to burnout organics, 124
segment 7—fourth heating segment, 124
segment 8—dwell at process temperature, 125
segment 9—crash cooling, 128
table summary of segments, 131

kiln procedures, 121
considerations in, 121
firing a casting, 122

kiln sitters, 132

kilns
definition of, 169
protection of during frit coloring, 113
safe positioning of, 155
safety tips for their use, 156
type recommended for crucible work, 113
use in frit production, 112
using to melt wax out of molds, 105

Klyr-Fire
use as a binding agent, 117

Kugler, 29

latex
description of, 66
incompatibility with plasticine, 66
using to make master molds, 66

lead glasses
appeal in kiln casting, 29
composition of, 29
effect of reducing atmospheres on, 19
leachability of, 29
properties of, 29
reduction of, 29
types of, 29

leather strips
use in making coddles, 87

leatherworking tools
using with waxes, 52

Lenox lead crystal, 40
life modeling, 68
light gage metal
use in making coddles, 86

Lillie, 37
lime, 27
definition of, 170
linoleum
use in making coddles, 86

liquids
structure of, 26

liquidus
of waxes, 48

lost wax casting
definition of, 43

ludo
as a property modifier, 79

magnets
use in frit preparation, 12

master mold
definition of, 170

master molds
attributes of, 63
from alginate, 64
from clay, 64
from Ggelflex, 65
from latex, 66
from moulage, 68
from plaster, 63
from RTV, 64
materials for, 63
using to cast replicas, 70

mesh numbers
definition of, 170
definition of sizes, 137
explanation of, 30
table of, 30

microcrystalline wax, 44
types of, 44

microwaves
using to steam wax out of molds, 107

mixing attachments
use with investments, 85

model weight
using to determine amount of frit required, 111

modeling
with wax, 43

models
definition of, 43
gating of, 61
orientation in molds, 61
preparing for casting, 61
removal from molds, 19

mold construction problems, 101
design problems, 101
handling problems, 102
material problems, 101
reasons for, 101
repairing, 102

mold frames
adgustable wood frames, 87
alternate methods of construction, 86
coddles, 87
definition of, 13, 170
doing without them, 98
flasks, 87
glueing to palette, 14
reinforcement of, 14
sizing of, 14

Index

using cardboard for, 13
mold repairs
    broken tenons, 104
    burrs or flashing, 102
    cracks, 103
    missing pieces, 104
    poorly mixed molds, 104
    voids or incomplete fills, 102
molds
    air drying of, 19
    cautions in filling, 20
    choosing wall thickness of, 13
    construction of, 85
    construction of waste molds, 13
    construction without mold frames, 98
    cracking of, 19
    curing of, 18, 19
    definition of, 170
    drying of, 107
    external reinforcement of, 88
    failure of, 20
    filling bas relief ones, 12
    filling in kiln, 20
    internal reinforcement of, 88
    melting wax out of, 104
    pouring of, 17
    problem with glass on tops of, 20
    reinforcement of, 88
    removal of, 22
    scraping edges of, 18
    screeding of, 18
    sizing of, 13
mortar and pestle
    use in frit manufacture, 112
mortises
    carving in press molds, 93
    requirements on, 93
    use in press molds, 92
moto-tools
    use on waxes, 50
moulage
    description of, 68
    using to make master molds, 68
multi-part freehand molds, 98
    constructing parting surface for, 98
    construction of base for, 100
    construction of reservoir for, 100
    investment application for, 99
    mixing investment for, 99
    separating completed mold, 100
    using backhand flick in making, 99
multi-part molds, 89
    adding a reservoir, 89
    alignment of, 91
    basic two part mold, 90
    finding the parting line for, 90
    securing sections together, 91
muriatic acid, 152
natches
    definition of, 170
    use in aligning multi-part molds, 91

natural gas
    checking for leaks of, 156
negative space inclusions, 120
newspaper
    use in making free hand molds, 99
normal consistency
    of gypsum plasters, 75
nylons
    using to sand waxes, 50
O'Hommel 33, 29
oils
    use in shining castings, 152
olivine sand
    as a refractory, 79
    binding agents for, 77
opacity
    cause of, 11
    choice of, 117
    control of, 109
    definition of, 170
    effect of mass on, 117
orbital sanders, 148
organics
    burnout of, 60
    model ideas, 60
    modeling with, 59
    undesirable model features, 59
overglazes
    use in shining castings, 153
paints
    use in shining castings, 153
paletes
    preparation of, 13
palettes
    choice of, 12
    definition of, 170
    used for, 12
paraffin, 46
parting line
    for multi-part molds, 90
pastes
    application of, 118
    binding agents for, 116
    blending colors in, 117
    making color samples of, 118
    reason for, 116
    suggested consistency of, 116
pâte de verre
    definition of, 170
PBI
    use in high-temperature gloves, 115
peaking
    definition of, 170
    during investment mixing, 16
    preparation of, 16
Pemco Pb 83, 29
physically absorbed water, 121
pinching
    in clay modeling, 54
plastecine, 46
    definition of, 170

© 2000 James Kervin and Dan Fenton

plaster
- carving of, 57
- modeling with, 56
- readying models for investment, 59
- scraping linear models, 58
- sculpting of, 56
- setting of, 18
- turning of, 57
- using to make master molds, 63

plaster bandages
- use in modeling, 57
- use with moulage master molds, 70

plaster molds
- using to cast wax shapes, 47

plaster of Paris, 76

plastic buckets
- considerations for use of, 115

plastic range
- of waxes, 48

plasticity
- of clays, 53

plastics
- modeling with, 60

plate glass, 29
- fusing of, 29

poise
- definition of, 170

poorly mixed molds
- repair of, 104

pour density, 31
pour volume, 111

powders
- using to blend colored pastes, 117

power relays, 133
powered handsaws, 140

press molds
- construction of, 92
- description of, 92
- filling, 95
- gates for, 94
- glass leakage from, 95
- separating after firing, 95
- separating completed mold, 94
- use of mortises with, 92
- using weights with, 92

pressing
- in clay modeling, 55

pressure cookers
- using to steam wax out of molds, 107

primary clay
- definition of, 53

process phase, 122

process temperature
- effects of, 110
- factors in choosing, 125
- table of suggested ones, 126
- tradeoff with process time, 126

programmable controllers, 133
- features of, 133

property modifiers
- alumina-silicate fibers, 79
- china clay, 80
- Fiberfrax, 79
- fiberglass fibers, 79
- grog, 79
- kaolin clay, 80
- ludo, 79
- stainless steel fibers, 80
- types of, 79
- zirconia fibers, 80

pyrometers, 134

quartz
- definition of, 25
- properties of, 27

quartz inversion
- of silica-ased refractories, 78

ramp
- definition of, 170

rasps
- use in shaping plaster

rate control
- in programmable controllers, 133

refractories
- alumina hydrate, 78
- crystobalite, 78
- definition of, 170
- diatomite, 78
- dry clay, 78
- kaolin clay, 78
- olivine sand, 79
- reasons for in investments, 77
- silica-based, 78
- types of, 77
- zirconia, 79

replica casting, 63
- finishing replicas, 72
- from life models, 68
- from master molds, 70
- hollow replicas, 72
- into alginate master molds, 71
- into latex master molds, 71
- into plaster master molds, 71
- into RTV master molds, 71

reservoirs
- definition of, 170
- sizing of, 61

respirators, 162
- care of, 164
- choosing, 163
- components of, 163
- consumption of, 164
- fitting, 163
- OSHA requirements, 163
- types of, 163

roofing paper
- use in making coddles, 86

RTV
- description of, 64
- using to make master molds, 64

safety, 155
- around rotating equipment, 155
- avoiding chemical burns, 158

Index

avoiding cuts, 155
avoiding fires, 155
avoiding thermal burns, 157
clothing recommendations for kiln work, 114
fires, 157
hazards of investments, 80
in crucible handling, 114
in frit manufacture, 112
in melting waxes, 47
in mixing large investment batches, 86
in mold removal, 22
in sawing, 140
in use of spindle grinders, 143
in using disk grinders, 143
in using toxic materials, 160
in wax melt out, 105
optical, 164
sand
   using to hold master molds, 71
   using to support molds, 20
sanding
   of waxes, 50
sanding equipment, 147
   belt sanders, 147
   disk sanders, 147
   drum sanders, 148
   hand sanding, 149
   orbital sanders, 148
   use of, 149
saw table guide, 140
sawing
   of waxes, 49
   safety considerations, 140
saws
   band saws, 138
   cutoff saws, 139
   handsaws, 140
   powered handsaws, 140
   table saws, 139
   types of, 138
   use of, 140
scraping
   linear plaster models, 58
screeding
   definition of, 18, 170
   fixing improper, 88
screens
   use as inclusions, 120
sculpting
   of waxes, 51
secondary clay
   definition of, 53
separators
   choices of, 15
   definition of, 170
set point controllers, 132
shellac
   use in preparing clay models, 56
   use in readying plaster models, 59
   use with moulage, 68
shims
   making, 98
   use in constructing multi-part freehand molds, 98
shock, 158
   symptoms of, 158
shrinkage
   in casting, 12
   planning for, 13
   using to tell firing progress, 21
silica flour
   as a refractory, 78
silica sand
   as a refractory, 78
silica tetrahedron, 26
silica-based refractories
   dangers of, 78
   quartz inversion of, 78
silicon carbide
   embedded tools, 138
   properties of, 138
silicosis, 78
slabbing
   in clay modeling, 55
   use in making press molds, 93
slaking
   definition of, 16, 170
soda
   definition of, 170
soda ash, 27
soda straws
   using to cast wax wires, 48
soda-lime glasses
   compatible lines of, 28
   composition of, 27
   definition of, 170
   effect of alumina additions, 28
   properties of, 27
sodium silicate
   use as a binding agent, 117
soft glass
   definition of, 170
softening point
   definition of, 170
   explanation of, 32
   using to determine annealing point, 37
soldering irons
   using on waxes, 51
solidus
   of waxes, 48
Spectrum Glass, 28
spindle grinders, 143
   construction of, 143
   safe use of, 143
   types of, 143
   use of, 143
sprues
   attaching to models, 61
squeegee oil
   use as a binding agent, 116
stainless steel fibers
   as a property modifier, 80
steam

effect on molds during curing, 19
steam injection
    using to steam wax out of molds, 106
stirring
    effect on investment cure, 16
    effect on set time, 17
    using to remove bubbles from investment, 17
strain
    definition of, 170
strain point
    definition of, 31, 170
    studio determination of, 40
stress
    definition of, 170
structures
    crystalline, 26
    liquid, 26
styrofoam
    modeling with, 60
Sure Form tools
    use in carving plaster, 57
    using to fix improper screeding, 88
surface texture
    during firing, 21
surface to volume ratios
    effect of allowable cooling rates, 39
table saws, 139
    care of, 139
    use of, 139
tap density, 31
    definition of, 170
tap volume, 111
temperature regimes, 31
    brittle solid regime, 31
    flexible solid regime, 32
    fluid regime, 32
    non-brittle solid regime, 31
tenons
    repairing brokens ones, 104
tension
    definition of, 170
terry cloth
    use in high-temperature gloves, 115
tetrahedron
    definition of, 170
texture
    control of, 110
texturing
    of waxes, 52
thermal shock
    definition of, 170
    use in frit manufacture, 112
thermocouples
    calibration of, 135
    factor affecting response, 134
    theory behind, 134
    types of, 135
tongs
    use in crucible handling, 113
tools
    cleaning of ones for investment, 18

toxic materials, 159
    chronic exposures to, 162
    degrees of, 160
    safe handling of, 160
    types of, 159
    types of inhalation hazards, 162
    use of respirators for, 162
    ventilation for, 161
translucency
    contolling in blending colored pastes, 117
    control of, 109
    definition of, 170
turning
    of plaster models, 57
Ullmann, 29
Ultracal cenent, 77
ultraviolet radiation, 165
undercuts
    definition of, 18
Uroboros Glass, 28
    available frit sizes of, 116
UV
    definition of, 170
vee gum
    use as a binding agent, 116
ventilation, 161
    design considerations for, 162
    types of, 162
vents
    for molds, 61
    in hollow vessel molds, 97
vibration
    using to remove bubbles, 17
victory wax, 44
viscosity
    definition of, 170
visible light, 165
voids or incomplete fills
    repair of, 102
volume multipliers
    using to determine amount of frit required, 111
Wasser Glass, 28
water
    types of in investment molds, 121
water displacement
    using to measure frit volume, 111
water displacement volume, 111
wax
    safe heating of, 156
wax extruder
    usig to cast replicas, 70
waxes
    addition of, 50
    beeswax, 46
    bonding of, 51
    bonding with sticky wax, 45
    burn out of, 19
    carving, 45
    carving of, 48, 49
    carving tools for, 49
    casting, 45

casting hollow shapes of, 48
casting shapes, 47
cleaning out of carpet, 47
compounded, 43
extruding wires of, 48
Ferris carving, 45
Ferris setup, 45
filing of, 50
flame polishing of, 52
French, 44
handworking of, 51
heating of, 50
history of, 43
liquidus, 48
melting, 46
melting out of molds, 104
microcrystalline, 44
microwaving out of molds, 107
mixing of, 44
modified, 43
modifying, 46
mold-a-wax, 45
natural, 43
paraffin, 46
plastic range of, 48
red, 44
removal from a mold, 19
reuse of, 44
safety in melting, 47
safety issues in melt out, 105
sanding of, 50
sawing of, 49
sculpting of, 51
setup, 45
shapes of, 44
solidus, 48
specifying properties of, 44
steaming out of molds, 106
sticky wax, 45
supporting with bench peg, 49
texturing of, 52
types of, 43
use in shining castings, 152
using moto-tools on, 50
victory, 44
warming of, 51
water soluble, 45
working with, 46
weights
    using with press molds, 92
wicket, 135
wire
    using for external reinforcement of molds, 88
    using to repair cracked molds, 103
wire staples
    installation of, 103
    using to repair cracked molds, 103
working point
    definition of, 170
    explanation of, 32
X-Acto knife
    using to carve waxes, 49
Zetex
    use in high-temperature gloves, 114
Zimmerman, 29
zirconia
    as a refractory, 79
zirconia fibers
    as a property modifier, 80

© 2000 James Kervin and Dan Fenton

If you enjoyed this book, you may be interested in Jim Kervin's book on glass beadmaking!!!

More Than You Ever Wanted To Know About Glass Beadmaking

With that book, you can master the many different techniques of glass beadmaking

You will discover how easy it is to:

Wind beads — Using the techniques of lampworking, you will learn to melt glass in a torch flame and wind it around a mandrel to form beads. You will then embellish these beads with the many decorations that you will learn to make at the torch.

Blow beads — You will learn to blow tubing to form lightweight hollow beads that can be shaped and decorated. These beads can even be made to contain small objects.

Fuse beads — Using a kiln, you will fuse sheet glass together to form pendant and tubular beads. Fine granules of glass will be fused together in a mold to make Pâte de Verre beads.

Draw beads — Using the techniques of glass blowing, you will learn how you can blow bubbles and then stretch them out to make bead tube stock. Decorations can be added by adorning the bubbles.

Press beads — You will see how equipment can be made to squish molten glass into shaped beads. Removable jaws allow you to make different sized and shaped beads.

Kiln cast beads — Using techniques similar to the ones in this book you will learn to make pâte de verre beads that are very unique.

Decorate beads — You will learn many different bead decoration techniques including: frits, dots, trailing, distortion, canes, latticino, ribbons, murrine, and much much more. The book also has a whole chapter devoted to the technique of making mosaic canes.

> **This is the most complete manual on glass beadmaking available. It is a must for anyone seriously interested in making glass beads.**

You will also learn to make other lampworked glass objects like marbles, cabochons, pendants, buttons, paper weights, and vessels both core and blown.

Listen to what others have said about that book:

Glass Line — *"As far as the reviewer is concerned this is the best book on glass beadmaking yet published and he is half tempted to apply the adjective "encyclopedic" to it."…"If you have an old edition, you should get a copy of the new one and make a gift of the old one to a novice glass beadmaker."*

Stained Glass — *"... the book could be titled, All That You Should Know About Glass"* ... *"I had the feeling as I was reading the book that someone was standing over my shoulder guiding me - a real bonus for a beginner."*

Common Ground Glass — *"The book is more than a lighthearted romp through the world of glass beadmaking, and goes beyond the capabilities of any video. If you are serious about beadmaking, you will want to add this manual to your reference collection."*

Glass Patterns Quarterly — *"Do you have questions on beadmaking? I'm betting you will find the answer in …"* (this book)

Lapidary Journal — *"We wanted to know! And you'll find out"*

Contemporary Lampworking — *"extensive text on making beads"*

The Glass Library — *"How did he do it? What a book!"*

Wale Apparatus Co., Inc. — *"This is a must"*

Frantz Bead Company — *"I highly recommend this book."*

Completely revised 4th edition has 294 pages, 371 figures, 22 tables, and 24 color plates with photographs of more than 100 different artists' work, even so this is no picture book — it is all information. Whether you are a veteran glass beadmaker seeking specific information or a newcomer looking for the basics, this book answers all your questions. It also makes a wonderful instructional text for those of you teaching glass beadmaking. It is the only book that adequately covers beadmaking equipment and safety. Recognized by NIOSH as an excellent source of safety knowledge.

Order your own copy of **More Than You Ever Wanted To Know About Glass Beadmaking** today. It is only $45, which includes shipping and handling. Send your check to:

> GlassWear Studios
> 1197 Sherry Way
> Livermore, CA 94550-5745
> or call (925)-443-9139
> or Fax (925)292-8648
> or E-mail jimkervin@compuserve.com

(ISBN 0-9651458-2-4)